Differentiation and integration

Mathematics for Engineers

The series is designed to provide engineering students in colleges and universities with a mathematical toolkit, each book including the mathematics in an engineering context. Numerous worked examples, and problems with answers, are included.

1. Laplace and z-transforms
2. Ordinary differential equations
3. Complex numbers
4. Fourier series
5. Differentiation and integration
6. Linear equations and matrices

Mathematics for Engineers

Differentiation and integration

W. Bolton

Routledge
Taylor & Francis Group

LONDON AND NEW YORK

First published 1995 by Longman Scientific & Technical

Published 2014 by Routledge
2 Park Square, Milton Park, Abingdon, Oxon OX14 4RN
711 Third Avenue, New York, NY 10017, USA

Routledge is an imprint of the Taylor & Francis Group, an informa business

Copyright © 1995, Taylor & Francis.

ISBN 13: 978-0-582-25180-9 (pbk)

British Library Cataloguing in Publication Data
A catalogue entry for this title is available from the British Library

Contents

Preface

This is one of the books in a series designed to provide engineering students in colleges and universities with a mathematical toolkit. In the United Kingdom it is aimed primarily at HNC/HND students and first-year undergraduates. Thus the mathematics assumed is that in BTEC National Certificates and Diplomas, the GNVQ Advanced level or in A level. The pace of development of the mathematics has been aimed at the notional reader for whom mathematics is not their prime interest or 'best subject' but need the mathematics in their other studies. The mathematics is developed and applied in an engineering context with large numbers of worked examples and problems, all with answers being supplied.

This book is concerned with the principles of differentiation and integration. A familarity with basic algebra and a basic knowledge of common functions, such as polynomials, trigonometric, exponential, logarithmic and hyperbolic, is assumed. An Appendix giving supportive mathematics on these functions is included. The aim of the book has been to include sufficient worked examples and problems to enable the reader to acquire some understanding and proficiency in the use of differentiation and integration to solve engineering problems.

W. Bolton

1 Introducing differentiation

1.1 Differentiation

Differentiation is a mathematical technique which is used to determine the rate at which functions change. There are many situations in engineering where we are concerned with rates of change. Thus we might have a current in an electrical circuit. Differentiation of the equation describing how the current varies with respect to time enables the rate at which the current varies with time to be established. We might have an object moving with a velocity which varies with time. Differentiation of the equation describing how the velocity varies with time enables the rate at which the velocity varies with time, i.e. the acceleration, to be determined.

This chapter introduces the basic concept of differentiation. Chapter 2 gives more details of the techniques of differentiation with chapter 3 illustrating the concept and techniques with a consideration of applications of differentiation in engineering. Before starting the discussion of differentiation in section 1.2, two key terms are briefly reviewed in this section, namely functions and limits. These terms will be found to be used frequently through the entire book.

1.1.1 Functions

In this, and later chapters, the term *function* is often used (see the Appendix). This term is used to indicate that there is some relationship between two quantities, without being specific about what the relationship is. If two variables are related such that the value of one depends on the value of the other, then one is said to be a function of the other. Thus, for example, for a car travelling along a road the distance s covered is a function of the time t it has been travelling. We can indicate this by writing $s = f(t)$. This expression just tells us that there is some relationship between the distance travelled and the time. The equation defining the relationship between s and t might be, for example, $s = 20t + 4t^2$.

1

As another example, the charge q on a capacitor is a function of the voltage V across it, thus we can write $q = f(V)$. The equation defining the relationship is $q = CV$, where C is a constant called the capacitance.

If $y = f(x)$ then $f(a)$ denotes the value of y when $x = a$. Thus, for example, $f(0)$ is the value of y when $x = 0$ and $f(2)$ is the value of y when $x = 2$. Once a function $f(x)$ has been defined by an equation then the value of $f(a)$ can be computed by substituting a for x in the equation used to define the function.

Example

If $y = x^2 + 2$, what are the values of (a) $f(0)$ and (b) $f(1)$?

(a) $f(0)$ is the value of the function when $x = 0$ and is thus 2.
(b) $f(1)$ is the value of the function when $x = 1$ and is thus 3.

Review problems

1   If $y = 3x + 4$, what are the values of (a) $f(0)$ and (b) $f(2)$?
2   The current i in a circuit is a function of time t and is described by the equation $i = 2(1 - e^{-5t})$ A. What is the value of $f(0)$?

1.1.2 Limits

The concept of a *limit* is an important feature of differentiation. Consider the following data which describe how the value of y depends on x when $y = x^2 + 2$:

$y = f(x)$	2.04	2.16	2.36	2.64	2.81	2.98
x	0.2	0.4	0.6	0.8	0.9	0.99

As the value of x gets closer and closer to 1 then the value of y gets closer and closer to 3.00. We can denote this as: as x tends to the value 1 then y tends to the value 3.00. Figure 1.1 illustrates this on a graph. In general we say that as x tends to 1 then y tends to a limit, i.e. the value 3.00. However close we make x to the value 1 it does not exceed the value 3.00. We write $x \rightarrow 1$ to denote that x tends to the value 1 and thus we can write

$$\lim_{x \to 1} f(x) = 3.00$$

This reads as: the limiting value of the function of x as x tends to the value 1 is 3.00.

As an engineering example, consider the current i in a circuit involving capacitance C in series with resistance R when a switch is closed and a steady voltage V is applied. The current is a

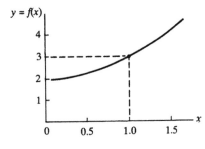

Fig. 1.1 $y = x^2 + 2$

function of time t, i.e. we can write $i = f(t)$, and is given by the equation

$$i = \frac{V}{R}e^{-t/RC}$$

As $t \to \infty$ then the exponential term tends to $e^{-\infty}$, i.e. 0. Thus

$$\lim_{t \to \infty} f(t) = 0$$

i.e. the current after an infinite amount of time, the so-called steady-state current which the circuit settles down to eventually, is zero.

Example

Determine the limiting value of $(4x - 1)$ as $x \to 1$.

As x tends to the value 1 then the function $(4x - 1)$ tends to the value 3.

Review problems

3 Determine:

(a) $\lim_{x \to 0} (2x^2 + 5)$, (b) $\lim_{x \to 2} (3x + 4)$, (c) $\lim_{\theta \to 0} \frac{\sin \theta}{\theta}$

(Hint: θ is in radians and so try values of $\theta = 0.5$, 0.1 and 0.01 to arrive at a sequence of values enabling you to deduce the value.)

1.2 Rate of change

Consider a quantity which varies with time and gives the graph shown in figure 1.2. It might, for example, be a distance–time graph for a moving object or the current–time graph for the current in an electrical circuit. At time t_1 the quantity has the value y_1. At a later time t_2 it has the value y_2. Thus y changes by the amount $y_2 - y_1$ in the time interval $t_2 - t_1$. The average rate of change of the quantity over this time interval is thus

$$\text{average rate of change} = \frac{y_2 - y_1}{t_2 - t_1} \qquad [1]$$

or BC/AC. The average rate of change is just the *gradient* (or *slope*) of the straight line joining A and B. The average rate of change of y with respect to t between two points is just the gradient of the straight line joining those two points.

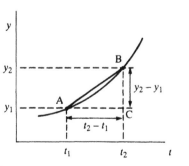

Fig. 1.2 Average rate of change

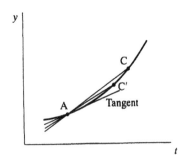

Fig. 1.3 Rate of change at point A

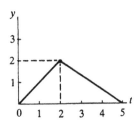

Fig. 1.4 Problem 5

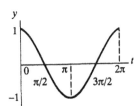

Fig. 1.5 Problem 7

The rate of change of y with respect to t is positive if when t increases in going from t_1 to t_2 we have y increasing in going from y_1 to y_2. The rate of change is negative if when t increases in going from t_1 to t_2 we have y decreasing in going from y_1 to y_2.

If we take the points A and C on the graph in figure 1.2 and move them closer and closer together, then we would reach a point when A and C coincide where we could consider we have the rate of change of y with t at a point. The straight line then becomes a *tangent* to the graph at that point. Figure 1.3 illustrates this. Thus the rate of change of y with respect to t at a point is the gradient of the tangent to the graph at that point.

Review problems

4 By sketching the graph of $y = 4x^2 - 2$, determine whether the rate of change of y with x is positive, negative or zero when (a) $x = 0$, (b) $x = -1$, (c) $x = +1$.

5 For the graph given in figure 1.4, estimate the rates of change of the function y with time t at (a) $t = 1$, (b) $t = 3$.

6 For the function $y = f(x)$ we have the relationship $y = x^2 + 2$. Sketch the graph of y against x and determine the rate of change of y with x at (a) $x = 0$, (b) $x = 1$.

7 For the graph of $y = \cos x$ shown in figure 1.5, determine the values of x at which the rate of change of y with x is (a) zero, (b) a maximum.

1.2.1 Velocity and acceleration

Velocity is defined as the rate of change of displacement with time, i.e. the rate of change of distance in a straight line with time. Thus the average velocity over a time interval t_1 to t_2 is the change in displacement that occurs in that time interval, say x_1 to x_2, divided by that time interval, i.e.

$$\text{average velocity} = \frac{x_2 - x_1}{t_2 - t_1}$$

The average velocity is positive if, for t_2 greater than t_1, we have x_2 greater than x_1. The velocity is negative if x_2 is less than x_1. The velocity at an instant of time is the value of the average velocity as $(t_2 - t_1)$ is made smaller and smaller and eventually tends to a zero value. It is thus the tangent to the displacement–time graph at that instant of time.

Acceleration is defined as the rate of change of velocity with time. Thus the average acceleration over a time interval t_1 to t_2 is the change in velocity that occurs in that time interval, say v_1 to v_2, divided by that time interval, i.e.

$$\text{average acceleration} = \frac{v_2 - v_1}{t_2 - t_1}$$

The average acceleration is positive if, for t_2 greater than t_1, we have v_2 greater than v_1. The acceleration is negative if v_2 is less than v_1. The acceleration at an instant of time is the value of the average acceleration as $(t_2 - t_1) \to 0$. It is thus the tangent to the velocity–time graph at that instant of time.

Review problems

8   The following data give the displacement of a particle at different times. What is the average velocity over the time interval (a) $t = 0$ s to $t = 3$ s, (b) $t = 1$ s to $t = 4$ s?

Distance in m	2	3	5	7	9
Time t in s	0	1	2	3	4

1.3 Derivatives

The average value of a rate of change tends to the instantaneous value of that rate of change at a point as the interval over which it is considered becomes smaller and smaller and eventually tends to zero. Thus, for example, for a velocity we have

$$\text{average velocity over a time interval} = \frac{x_2 - x_1}{t_2 - t_1}$$

where x_1 is the displacement at time t_1 and x_2 the displacement at time t_2. We can write this difference in displacement as δx and this difference in time as δt. The δ symbol in front of a quantity means 'a small bit of it' or 'an interval of '. Thus the equation can be written as

$$\text{average velocity over a time interval} = \frac{\delta x}{\delta t}$$

An alternative symbol which is often used is Δx, with the Δ symbol being used to indicate that we are referring to a small bit of the quantity x. Either of these forms of notation should not be considered as meaning that we have δ or Δ multiplying x. The δx or Δx should be considered as a single symbol representing a single quantity.

As we make the time interval δt smaller, the average velocity becomes closer to the velocity at an instant of time. Eventually when the time interval tends to zero then we have the velocity at an instant of time. We can write this as

$$\lim_{\delta t \to 0} \frac{\delta x}{\delta t} = \frac{dx}{dt}$$

This reads as: the limiting value of $\delta x/\delta t$ as δt tends to a zero value equals dx/dt. A *limit* is a value to which we get closer and closer as we carry out some operation. Thus dx/dt is the value of the rate of change at an instant of time. For a graph of displacement against time then dx/dt is the tangent to a point on the graph; dx/dt is called the *derivative* of a function at a particular point and is read as 'the derivative of x with respect to t'. The process of determining the derivative for a function is called *differentiation*. The notation dx/dt should not be considered as d multiplied by x divided by d multiplied by t, but as a single symbol representing the rate of change of x with t.

Thus if we have a displacement x which is a function of time t then we can write $x = f(t)$. The value of x at some value of t is represented by $f(t)$. The value of x at $t + \delta t$ is written as $f(t + \delta t)$. The derivative of the function at some value of t is thus

$$\frac{dx}{dt} = \lim_{\delta t \to 0} \frac{f(t + \delta t) - f(t)}{\delta t} \qquad [2]$$

There are a number of other forms of notation that are used to represent derivatives, namely

$$\frac{d}{dt}\{f(t)\} \quad \text{or} \quad \frac{dx}{dt} \quad \text{or} \quad f'(t) \quad \text{or} \quad x' \quad \text{or} \quad D(x)$$

1.3.1 Determining derivatives

Suppose we have y which is a function of x and given by the equation $y = x^2$. The derivative of this function is given by equation [2] as

$$\frac{dy}{dx} = \lim_{\delta x \to 0} \frac{f(x + \delta x) - f(x)}{\delta x}$$

The value of the function at x is x^2. The value of the function at $x + \delta x$ is $(x + \delta x)^2$. Thus

$$\frac{dy}{dx} = \lim_{\delta x \to 0} \frac{(x + \delta x)^2 - x^2}{\delta x}$$

$$= \lim_{\delta x \to 0} \frac{x^2 + 2x\,\delta x + (\delta x)^2 - x^2}{\delta x}$$

$$= \lim_{\delta x \to 0} (2x + \delta x)$$

As δx tends to 0 then we end up with

$$\frac{dy}{dx} = 2x$$

or, written in another way,

$$\frac{d}{dx}(x^2) = 2x$$

Example

The displacement x of an object varies with time t according to the equation $x = 2t + 3t^2$. Determine the derivative of the function and the value of the rate of change of x with t at $t = 1$.

Using equation [2],

$$\frac{dx}{dt} = \lim_{\delta t \to 0} \frac{f(t + \delta t) - f(t)}{\delta t}$$

$$= \lim_{\delta t \to 0} \frac{\{2(t + \delta t) + 3(t + \delta t)^2\} - \{2t + 3t^2\}}{\delta t}$$

$$= \lim_{\delta t \to 0} (2 + 6t + 3\delta t)$$

$$= 2 + 6t$$

The above equation indicates that the gradient of the function is also a function of t. At $t = 1$, the rate of change of x with t will have the value $2 + 6 = 8$.

Review problems

9   What are the values of the gradients of the following functions at $x = 0$ and $x = 1$:

(a) $y = 2x^2 + 3x$, (b) $y = x^2 + 2x + 3$, (c) $y = 2x + 1$?

1.3.2 Existence of derivatives

We can interpret the derivative as representing the gradient of the tangent at a particular point on a graph of the function concerned. This means that with a continuous function, e.g. a function $y = f(x)$ which has values of y which smoothly and continuously change as x changes, we have derivatives for all values of x. However, with a discontinuous graph this is not true. For such functions there are some values of x for which there is no derivative. For example, for the function shown in figure 1.6(a) there is no gradient to the

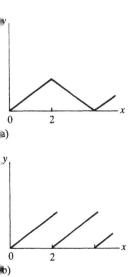

a)

b)

Fig. 1.6 Discontinuous graphs

graph at $x = 2$ since the graph is a point at that value. For the function shown in figure 1.6(b) there is no value to the gradient at the break in the graph at $x = 2$.

The following are points on graphs at which there would be no derivative:

1 Discontinuities
2 Sharp turns
3 Vertical lines

Review problems

10 By sketching graphs of the following functions, determine the value of x at which the derivative is undefined:
(a) $y = 1/x$,
(b) $y = x^2 + 1$ for values of x less than 0 and $y = -x^2$ for values of x greater than 0,
(c) $y = 1$ for values of x greater than 1 and $y = 0$ for values of x less than 1 (this is called a unit step function).

1.4 Common derivatives

Table 1.1 lists some commonly encountered functions and their derivatives. These functions can be combined in a variety of ways to give functions encountered in engineering. Chapter 2 is about combining such functions. The functions can then be directly used, as for example in chapter 3, to solve problems involving derivatives without the need to derive them from first principles. The following sections illustrate, by considering a few examples, how the derivatives in the table were obtained from first principles, i.e. using equation [2].

1.4.1 $y = x^n$

If $y = x^n$ then if we consider a small increase in x of δx the corresponding increase in the value of y is δy where

$$y + \delta y = (x + \delta x)^n$$

Expanding this by the use of the binomial theorem gives

$$y + dy = x^n + nx^{n-1}\,\delta x + \frac{n(n-1)}{2!}x^{n-2}(\delta x)^2 + \ldots + (\delta x)^n$$

Thus, using equation [2],

$$\frac{dy}{dx} = \lim_{\delta x \to 0} \frac{f(x + \delta x) - f(x)}{\delta x}$$

Table 1.1 Derivatives

	Function	Derivative
1	a constant	0
2	x^n	nx^{n-1}
3	e^{at}	$a\,e^{at}$
4	$\ln x$	$1/x$
5	$\sin(ax+b)$	$a\cos(ax+b)$
6	$\cos(ax+b)$	$-a\sin(ax+b)$
7	$\tan(ax+b)$	$a\sec^2(ax+b)$
8	$\sec(ax+b)$	$a\sec(ax+b)\tan(ax+b)$
9	$\operatorname{cosec}(ax+b)$	$-a\operatorname{cosec}(ax+b)\cot(ax+b)$
10	$\cot(ax+b)$	$-a\operatorname{cosec}^2(ax+b)$
11	$\sin^{-1}(ax+b)$	$\dfrac{a}{\sqrt{1-(ax+b)^2}}$
12	$\cos^{-1}(ax+b)$	$-\dfrac{a}{\sqrt{1-(ax+b)^2}}$
13	$\tan^{-1}(ax+b)$	$\dfrac{a}{1+(ax+b)^2}$
14	$\sinh(ax+b)$	$a\cosh(ax+b)$
15	$\cosh(ax+b)$	$a\sinh(ax+b)$
16	$\tanh(ax+b)$	$a\operatorname{sech}^2(ax+b)$
17	$\operatorname{sech}(ax+b)$	$-a\operatorname{sech}(ax+b)\tanh(ax+b)$
18	$\operatorname{cosech}(ax+b)$	$-a\operatorname{cosech}(ax+b)\coth(ax+b)$
19	$\coth(ax+b)$	$-a\operatorname{cosech}^2(ax+b)$
20	$\sinh^{-1}(ax+b)$	$\dfrac{a}{\sqrt{(ax+b)^2+1}}$
21	$\cosh^{-1}(ax+b)$	$\dfrac{a}{\sqrt{(ax+b)^2-1}}$
22	$\tanh^{-1}(ax+b)$	$\dfrac{a}{1-(ax+b)^2}$

$$= \lim_{\delta x \to 0} \left\{ nx^{n-1} + \frac{n(n-1)}{2!}x^{n-2}\delta x + \ldots + (dx)^{n-1} \right\}$$

and so we have

$$\frac{\mathrm{d}}{\mathrm{d}x}(x^n) = nx^{n-1} \tag{3}$$

This relationship applies for positive, negative and fractional values of n.

If $y = Cx^n$, where C is a constant, then equation [2] gives

$$\frac{dy}{dx} = \lim_{\delta x \to 0} \frac{C(x + \delta x)^n - Cx}{\delta x}$$

$$= \lim_{\delta x \to 0} C \frac{(x + \delta x)^n - x^n}{\delta x}$$

$$= C \lim_{\delta x \to 0} \frac{(x + \delta x)^n - x}{\delta x}$$

and so we have

$$\frac{d}{dx}(Cx^n) = Cnx^{n-1} \tag{4}$$

Note that if we have just $y = C$ that equation [2] gives

$$\frac{dy}{dx} = \lim_{\delta x \to 0} \frac{C - C}{\delta x} = 0 \tag{5}$$

The equation $y = C$ represents a horizontal straight line and the gradient of it is 0.

Example

Determine the derivative of the function $y = x^5$.

Using equation [3],

$$\frac{dy}{dx} = 5x^4$$

Example

Determine the derivative of the function $y = 1/x$.

This function can be written as $y = x^{-1}$. Thus, using equation [3],

$$\frac{dy}{dx} = -1x^{-2} = -\frac{1}{x^2}$$

Example

Determine the derivative of the function $y = 3x^{3/4}$.

Using equation [3],

$$\frac{dy}{dx} = 3 \times \frac{3}{4} x^{-1/4}$$

Example

Determine the rate of change of volume of a cube with respect to its side length.

The volume V of a cube is given by $V = L^3$, where L is the length of a side. Thus

$$\frac{dV}{dL} = 3L^2$$

Review problems

11 Determine the derivatives of the following functions:
   (a) $y = x^3$, (b) $y = x^6$, (c) $y = x^{-2}$, (d) $y = x^{3/2}$, (e) $y = \sqrt{x}$,
   (f) $y = 2x^2$, (g) $y = 3x^{-1/2}$, (h) $y = 3$

12 Determine the rate at which a flywheel rotates, i.e. the angle θ varies with time t, if the angle is related to the time by the equation $\theta = 4t^2$.

13 What is the gradient of the graph of y against x at $x = 1$ if we have $y = 3x^2$?

14 Determine the rate of change of the area of a circle with respect to its radius when the radius is 20 mm.

1.4.2 Trigonometric functions

This section starts with a brief reiteration of the definitions and some of the basic relationships for trigonometric functions, before discussing differentiation. The six trigonometric functions are defined in terms of the angle θ in a right-angled triangle.

$$\sin\theta = \frac{\text{side opposite to angle}}{\text{hypotenuse}} \qquad [6]$$

$$\cos\theta = \frac{\text{side adjacent to angle}}{\text{hypotenuse}} \qquad [7]$$

$$\tan\theta = \frac{\text{side opposite to angle}}{\text{side adjacent to angle}} = \frac{\sin\theta}{\cos\theta} \qquad [8]$$

$$\sec\theta = \frac{1}{\cos\theta} \qquad [9]$$

$$\operatorname{cosec}\theta = \frac{1}{\sin\theta} \qquad [10]$$

$$\cot\theta = \frac{1}{\tan\theta} \qquad [11]$$

Since, by the use of the Pythagoras theorem,

$$(\text{hypotenuse})^2 = (\text{side opposite to angle})^2$$
$$+ (\text{side adjacent to angle})^2$$

then, dividing throughout by $(\text{hypotenuse})^2$ we have

$$1 = \sin^2\theta + \cos^2\theta \qquad [12]$$

If this equation is divided throughout by $\sin^2\theta$ we obtain

$$\mathrm{cosec}^2\theta = 1 + \cot^2\theta \qquad [13]$$

Dividing equation [12] by $\cos^2\theta$ gives

$$\sec^2\theta = \tan^2\theta + 1 \qquad [14]$$

It is often useful to be able to express the trigonometric functions of angles such as $A + B$ or $A - B$ in terms of the trigonometric functions of A and B. Such relationships include

$$\sin (A + B) = \sin A \cos B + \cos A \sin B \qquad [15]$$

$$\cos (A + B) = \cos A \cos B - \sin A \sin B \qquad [16]$$

These relationships can also be used to generate equations for double angles, i.e. when $A = B$. For example

$$\sin 2A = \sin A \cos A + \cos A \sin A = 2 \sin A \cos A \qquad [17]$$

$$\cos 2A = \cos^2 A - \sin^2 A \qquad [18]$$

The Appendix lists all the compound and double angle equations.
If $y = \sin x$ then if we consider a small increase in x of δx the corresponding increase in the value of y is δy where

$$y + \delta y = \sin(x + \delta x)$$

Thus

$$f(x + \delta x) - f(x) = \sin(x + \delta x) - \sin x$$

But $\sin A - \sin B = 2 \cos \frac{1}{2}(A + B) \sin \frac{1}{2}(A - B)$, thus

$$f(x + \delta x) - f(x) = 2 \cos\left(x + \frac{\delta x}{2}\right) \sin\left(\frac{\delta x}{2}\right)$$

and so, using equation [2],

$$\frac{dy}{dx} = \lim_{\delta x \to 0} \frac{f(x + \delta x) - f(x)}{\delta x}$$

$$= \lim_{\delta x \to 0} \frac{2 \cos\left(x + \frac{\delta x}{2}\right) \sin\left(\frac{\delta x}{2}\right)}{\delta x}$$

For small angles the sine approximates to the angle in radians, i.e. $\sin(\delta x/2)$ is approximately $\delta x/2$. Thus $[\sin(\delta x/2)]/(\delta x/2)$ tends to the value 1 as δx tends to 0. As δx tends to 0 then the cosine term will tend to $\cos x$. Thus

$$\frac{d}{dx}(\sin x) = \cos x \qquad\qquad [19]$$

Figure 1.7 shows a graph of the function and its derivative. The derivative has a zero value when the function has a maximum value, this being the point at which the gradient of $\sin x$ is zero. The derivative has a maximum value when the function has a zero value, this being the point at which the gradient of $\sin x$ is a maximum.

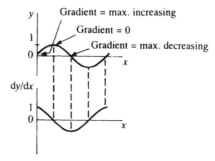

Fig. 1.7 (a) y = sin x,
(b) dy/dx = cos x

If we had considered $y = \sin ax$, where a is a constant, then we would have obtained

$$\frac{d}{dx}(\sin ax) = a \cos ax \qquad\qquad [20]$$

If $y = \cos x$ then if we consider a small increase in x of δx the corresponding increase in the value of y is δy where

$$y + \delta y = \cos(x + \delta x)$$

Thus

$$f(x + \delta x) - f(x) = \cos(x + \delta x) - \cos x$$

But $\cos A - \cos B = -2 \sin \frac{1}{2}(A + B) \sin \frac{1}{2}(A - B)$. Thus

$$f(x + \delta x) - f(x) = -2 \sin\left(x + \frac{\delta x}{2}\right) \sin\left(\frac{\delta x}{2}\right)$$

and so, using equation [2],

$$\frac{dy}{dx} = \lim_{\delta x \to 0} \frac{f(x + \delta x) - f(x)}{\delta x}$$

$$= \lim_{\delta x \to 0} \frac{-2 \sin\left(x + \frac{\delta x}{2}\right) \sin\left(\frac{\delta x}{2}\right)}{\delta x}$$

As before, $[\sin(\delta x/2)]/(\delta x/2)$ tends to 1 as δx tends to 0. As δx tends to 0 we also have $\sin(x + \delta x/2)$ tending to $\sin x$. Thus

$$\frac{d}{dx}(\cos x) = -\sin x \qquad\qquad [21]$$

If we had considered $y = \cos ax$, where a is a constant, then we would have obtained

$$\frac{d}{dx}(\cos ax) = -a \sin ax \qquad\qquad [22]$$

If $y = \tan x$ then if we consider a small increase in x of δx the corresponding increase in the value of y is δy where

$$y + \delta y = \tan(x + \delta x)$$

Thus

$$f(x + \delta x) - f(x) = \tan(x + \delta x) - \tan x$$

$$= \frac{\sin(x + \delta x)}{\cos(x + \delta x)} - \frac{\sin x}{\cos x}$$

$$= \frac{\sin(x + \delta x)\cos x - \sin x \cos(x + \delta x)}{\cos(x + \delta x)\cos x}$$

Since $\sin(A - B) = \sin A \cos B - \cos A \sin B$, then we can write

$$f(x + \delta x) - f(x) = \frac{\sin(x + \delta x - x)}{\cos(x + \delta x)\cos x}$$

Thus, using equation [2],

$$\frac{dy}{dx} = \lim_{\delta x \to 0} \frac{f(x + \delta x) - f(x)}{\delta x}$$

$$= \lim_{\delta x \to 0} \frac{\sin \delta x}{\delta x} \frac{1}{\cos(x + \delta x)\cos x}$$

Since, as before, when $\delta x \to 0$ then $(\sin \delta x)/\delta x \to 1$. Hence

$$\frac{d}{dx}(\tan x) = \frac{1}{\cos^2 x} = \sec^2 x \qquad [23]$$

If we had considered $y = \tan ax$, where a is a constant, then we would have obtained

$$\frac{d}{dx}(\tan ax) = a \sec^2 ax \qquad [24]$$

The derivatives of cosec θ, sec θ and cot θ can be obtained in a similar manner; however, it is fairly easy to obtain them using the above results for sine, cosine and tangent with the quotient rule discussed in chapter 2. For completeness the results are quoted here.

$$\frac{d}{dx}(\text{cosec } ax) = -a \text{ cosec } ax \cot ax \qquad [25]$$

$$\frac{d}{dx}(\sec ax) = a \sec ax \tan ax \qquad [26]$$

$$\frac{d}{dx}(\cot ax) = -a \text{ cosec}^2 ax \qquad [27]$$

Example

Determine the derivative of the function $y = \sin 2x$.

Using equation [20],

$$\frac{dy}{dx} = 2 \cos 2x$$

Example

The current i in an electrical circuit varies with time t and is given by $i = \cos 100t$. Derive an equation for the rate of change of the current with time.

Using equation [22],

$$\frac{di}{dt} = -100 \sin 100t$$

Review problems

15 Determine the derivatives of the following functions:
(a) $y = \cos x$, (b) $y = \sin 5x$, (c) $y = \cos 4x$, (d) $y = \tan 2x$

16 The displacement x of an oscillating object varies with time t according to the equation $y = \cos 20t$. Derive an equation indicating how the velocity varies with time.

1.4.3 Exponential functions

If $y = e^x$ then if we consider a small increase in x of δx the corresponding increase in the value of y is δy where

$$y + \delta y = e^{x+\delta x} = e^x e^{\delta x}$$

Thus

$$f(x + \delta x) - f(x) = e^x e^{\delta x} - e^x = e^x(e^{\delta x} - 1)$$

We can, in general, express an exponential as a series

$$e^x = 1 + x + \frac{x^2}{2!} + \frac{x^3}{3!} + \ldots$$

Hence we can write

$$f(x + dx) - f(x) = e^x\left(\delta x + \frac{(\delta x)^2}{2!} + \frac{(\delta x)^3}{3!} + \ldots\right)$$

Thus, using equation [2],

$$\frac{dy}{dx} = \lim_{\delta x \to 0} \frac{f(x + \delta x) - f(x)}{\delta x}$$

$$= \lim_{\delta x \to 0} e^x\left(1 + \frac{\delta x}{2!} + \frac{(\delta x)^2}{3!} + \ldots\right)$$

and so

$$\frac{d}{dx}(e^x) = e^x \tag{28}$$

If we had $y = e^{ax}$ then

$$\frac{d}{dx}(e^{ax}) = a e^{ax} \tag{29}$$

Example

Determine the derivative of the function $y = e^{3x}$.

Using equation [29],

$$\frac{dy}{dx} = 3 e^{3x}$$

Review problems

17 Determine the derivatives of the following functions:
   (a) $y = e^{2x}$, (b) $y = e^{-4x}$, (c) $y = e^{3x/2}$
18 The number of radioactive atoms N in a radioactive element changes with time t according to the equation $N = N_0 e^{-\lambda t}$. Determine the rate of change of N with time (note that is called the activity).

1.4.4 Logarithmic functions

If $y = \ln x$, then we can write this as

$$x = e^y$$

Differentiating with respect to y, by the use of equation [28], gives

$$\frac{dx}{dy} = e^y = x$$

Hence, if we invert this we have

$$\frac{d}{dx}(\ln x) = \frac{1}{x} \qquad\qquad [30]$$

Example

Determine the derivative of $y = \ln 2x$.

We can rewrite the equation as

$$2x = e^y$$

Thus

$$2\frac{dx}{dy} = e^y = 2x$$

and so

$$\frac{dy}{dx} = \frac{1}{x}$$

Note that an alternative way of obtaining this answer would have been to write the initial equation as

$$y = \ln 2 + \ln x$$

Then, since the derivative of 2 with respect to x is 0 we just have to deal with $y = \ln x$.

Review problems

19 Determine the derivative of $y = 2 \ln x$.

1.4.5 Hyperbolic functions

This section starts with a brief discussion of hyperbolic functions before considering differentiation.

Hyperbolic functions are combinations of exponential functions and occur quite often in engineering, e.g. in transmission line theory. There are six hyperbolic functions, the six having properties similar to those of the six trigonometric functions. They are defined as follows:

$$\sinh x = \frac{e^x - e^{-x}}{2} \tag{31}$$

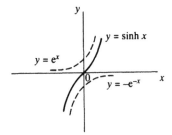

$$y = e^x$$

Fig. 1.8 sinh x

sinh x is pronounced as 'shine x'. Figure 1.8 shows a graph of sinh x plotted against x. As indicated by the above equation, the graph can be regarded as the average value of the graphs of e^x and $-e^{-x}$. From the graph we have sinh $0 = 0$ and $\sinh(-x) = -\sinh x$. Note that, unlike the trigonometric function of sin x, the sinh x is not a periodic graph.

$$\cosh x = \frac{e^x + e^{-x}}{2} \tag{32}$$

cosh x is pronounced as 'kosh x'. Figure 1.9 shows a graph of cosh x plotted against x. As equation [32] indicates, the graph can be regarded as the average of e^x and e^{-x}. From the graph we have $\cosh(0) = 1$ and $\cosh(-x) = \cosh x$. Note that, unlike the trigonometric function of cos x, the cosh x graph is not periodic. The shape of the curve is that of a heavy rope or chain between supports and hanging freely under gravity, e.g. a telephone cable between telegraph poles. The shape is called a *catenary*.

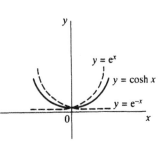

Fig. 1.9 cosh x

$$\tanh x = \frac{\sinh x}{\cosh x} = \frac{e^x - e^{-x}}{e^x + e^{-x}} \tag{33}$$

tanh x is pronounced as 'than x'. Figure 1.10 shows a graph of tanh x plotted against x. From the graph we have tanh$(0) = 0$,

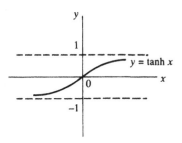

Fig. 1.10 tanh x

$\tanh(-x) = -\tanh x$, and as x tends to infinity then $\tanh x$ tends to the value of 1. Note that, unlike the trigonometric function of $\tan x$, the $\tanh x$ graph is not periodic.

$$\operatorname{cosech} x = \frac{1}{\sinh x} \qquad [34]$$

$\operatorname{cosech} x$ is pronounced as 'coshec x'.

$$\operatorname{sech} x = \frac{1}{\cosh x} \qquad [35]$$

$\operatorname{sech} x$ is pronounced as 'shec x'.

$$\coth x = \frac{1}{\tanh x} \qquad [36]$$

$\coth x$ is pronounced as 'koth x'.

We can obtain various relationships between the hyperbolic functions (see the Appendix). Thus, using equations [31] and [32],

$$\cosh x + \sinh x = \tfrac{1}{2}(e^x + e^{-x}) + \tfrac{1}{2}(e^x - e^{-x}) = e^x$$

$$\cosh x - \sinh x = \tfrac{1}{2}(e^x + e^{-x}) - \tfrac{1}{2}(e^x - e^{-x}) = e^{-x}$$

Hence

$$(\cosh x + \sinh x)(\cosh x - \sinh x) = e^x e^{-x} = e^0 = 1$$

Thus

$$\cosh^2 x - \sinh^2 x = 1 \qquad [37]$$

Dividing each term in this equation by $\cosh^2 x$ gives

$$1 - \tanh^2 x = \operatorname{sech}^2 x \qquad [38]$$

Dividing each term in equation [37] by $\sinh^2 x$ gives

$$\coth^2 x - 1 = \operatorname{cosech}^2 x \qquad [39]$$

We can likewise deduce relationships for compound angle addition and subtraction. For example we have

$$\sinh(A + B) = \sinh A \cosh B + \cosh A \sinh B \qquad [40]$$

$$\cosh(A + B) = \cosh A \cosh B + \sinh A \sinh B \qquad [41]$$

There is a rule, called *Osborne's rule*, which states that the six trigonometric functions used in trigonometric identities may be replaced by their corresponding hyperbolic functions provided the sign of any direct or implied product of two sines is changed. Thus, for example, we have

$$\cos^2 x + \sin^2 x = 1$$

$$\cosh^2 x - \sinh^2 x = 1$$

The Appendix gives a complete list of these compound hyperbolic function relationships and double hyperbolic function relationships.

Consider the differentiation of the function $y = \sinh x$. Using equation [15] we can write this as

$$\frac{d}{dx}(\sinh x) = \frac{d}{dx}\left(\frac{e^x - e^{-x}}{2}\right) = \frac{1}{2}e^x - (-\frac{1}{2}e^{-x}) = \cosh x \qquad [42]$$

With $y = \sinh ax$ we obtain

$$\frac{d}{dx}(\sinh ax) = a\cosh a \qquad [43]$$

Consider the differentiation of the function $y = \cosh x$. Using equation [16] we can write this as

$$\frac{d}{dx}(\cosh x) = \frac{d}{dx}\left(\frac{e^x + e^{-x}}{2}\right) = \frac{1}{2}e^x + (-\frac{1}{2}e^{-x}) = \sinh x \qquad [44]$$

With $y = \cosh ax$ we obtain

$$\frac{d}{dx}(\cosh ax) = a\sinh ax \qquad [45]$$

Consider the differentiation of the function $y = \tanh x$. For completeness this function is considered in this chapter, though the method used is not developed until chapter 2. Since we can write $\tanh x = \sinh x/\cosh x$ we can use the quotient rule (see chapter 2). Thus

$$\frac{d}{dx}(\tanh x) = \frac{\cosh x \cosh x - \sinh x \sinh x}{\cosh^2 x}$$

Using equation [21] we thus obtain

$$\frac{d}{dx}(\tanh x) = \frac{1}{\cosh^2 x} = \text{sech}^2 x \qquad [46]$$

With $y = \tanh ax$ we obtain

$$\frac{d}{dx}(\tanh ax) = a \operatorname{sech}^2 ax \qquad\qquad [47]$$

The derivatives of sech ax, cosech ax and coth ax can also be derived using this quotient rule. They are

$$\frac{d}{dx}(\operatorname{sech} ax) = -a \operatorname{sech} ax \tanh ax \qquad\qquad [48]$$

$$\frac{d}{dx}(\operatorname{cosech} ax) = -a \operatorname{cosech} ax \coth ax \qquad\qquad [49]$$

$$\frac{d}{dx}(\coth ax) = -a \operatorname{cosec}^2 ax \qquad\qquad [50]$$

Example

What is the value of sinh 4?

Using equation [31],

$$\sinh x = \frac{e^x - e^{-x}}{2} = \frac{e^4 - e^{-4}}{2} = \frac{54.598 - 0.018}{2} = 27.29$$

Example

Determine the derivatives of the following functions:
(a) $y = \sinh 2x$, (b) $y = \tanh 3x$

(a) Using equation [43],

$$\frac{d}{dx}(\sinh 2x) = 2 \cosh 2x$$

(b) Using equation [47],

$$\frac{d}{dx}(\tanh 3x) = 3 \operatorname{sech}^2 3x$$

Review problems

20 Determine the derivatives of the following functions:
(a) $y = \cosh 2x$, (b) $y = \sinh 5x$, (c) $y = \tanh 7x$

1.5 Further differentiation

The derivative dy/dx of the function $y = f(x)$ is itself a function and may also be differentiated. The derivative of a derivative is called the *second derivative* and can be written in a number of forms:

$$\frac{d}{dx}\left(\frac{dy}{dx}\right) \quad \text{or} \quad \frac{d^2y}{dx^2} \quad \text{or} \quad \frac{d^2}{dx^2}\{f(x)\} \quad \text{or} \quad f''(x) \quad \text{or} \quad D^2x$$

This may in turn be differentiated to give the third derivative, this being written as

$$\frac{d}{dx}\left(\frac{d^2y}{dx^2}\right) \quad \text{or} \quad \frac{d^3y}{dx^3} \quad \text{or} \quad \frac{d^3}{dx^3}\{f(x)\} \quad \text{or} \quad f'''(x) \quad \text{or} \quad D^3x$$

This may in turn be differentiated, and so on for higher derivatives.

The first derivative gives information about the tangents to a graph, a graph of the first derivative indicating how the gradients of the tangents change. The second derivative gives information about the rate of change of the gradient of the tangents. Thus, for example, for a moving object we can have a relationship between the displacement x of the object and time t. The first derivative of this relationship gives the rate of change of displacement with time, i.e. the velocity. If we differentiate the velocity with respect to time we obtain the rate of change of velocity with time, i.e. the acceleration. This is the second derivative of the displacement with time. Thus

$$\text{velocity } v = \frac{dx}{dt}$$

$$\text{acceleration } a = \frac{dv}{dt} = \frac{d^2x}{dt^2}$$

Example

Determine the second derivative of $y = x^3$.

The first derivative is

$$\frac{dy}{dx} = 3x^2$$

The second derivative is given by differentiating this equation again. Thus

$$\frac{d^2y}{dx^2} = 6x$$

Example

The displacement x of an object with time t is given by the equation $x = 4t^2$. Determine how the acceleration of the object changes with time.

The first derivative is

$$\frac{dx}{dt} = 8t$$

The acceleration is the second derivative and so is

$$\text{acceleration} = \frac{d^2x}{dt^2} = 8$$

Review problems

21 Determine the second derivatives of the following functions:
    (a) $y = 2x^2$, (b) $y = \sin x$, (c) $y = e^{2x}$, (d) $y = \ln x$
22 The charge q on the plates of a capacitor is related to time t by
    the equation $q = CV\, e^{-t/RC}$, where C, V and R are constants.
    Determine how (a) the current, i.e. dq/dt, changes with time
    and (b) how the rate of change of current with time, changes
    with time, i.e. the second derivative.

Further problems

23 Working from first principles, determine the derivatives of the
    following functions: (a) $y = \sqrt{x}$, (b) $y = 2/x$.
24 Determine, using table 1.1, the derivatives of the following
    functions:

    (a) $y = x^6$, (b) $y = x^{-2}$, (c) $y = x^{5/3}$, (d) $y = 2x^3$, (e) $y = 2\sqrt{x}$,

    (f) $y = 5x^{-3/2}$, (g) $y = 4$, (h) $y = \sin 4x$, (i) $y = \cos 2x$, (j) $y = e^x$,

    (k) $y = e^{2x}$, (l) $y = e^{-2x/3}$, (m) $y = \ln 4x$, (n) $y = \tan 5x$,

    (o) $y = \sin 5x$, (p) $y = \sinh 3x$, (q) $y = \tanh 2x$

25 The displacement s of an object starting from rest and moving
    with a uniform acceleration a is given by $s = \frac{1}{2}at^2$. Derive an
    equation for the velocity.
26 Determine the gradient of the graph of $y = \cos 2x$ when:
    (a) $x = \pi/2$, (b) $x = \pi/4$, (c) $x = 0$.
27 Determine the rate of change of area of a square with respect
    to the length of a side when the length is 20 mm.
28 The kinetic energy E of a rotating flywheel is a function of the
    angular velocity ω and given by $E = \frac{1}{2}I\omega^2$. Determine the rate
    of change of kinetic energy with respect to angular velocity.
29 The energy E stored by a capacitor is a function of the
    potential difference V across it, being given by $E = \frac{1}{2}CV^2$.

Determine the rate of change of energy with respect to potential difference.

30 The atmospheric pressure p depends on the height h above ground level at which the measurement is made, being related by the equation $p = p_0 e^{-ch}$, where p_0 and c are constants. Determine the rate of change of pressure with height.

31 Determine the second derivatives of the following functions:
(a) $y = 3x^4$, (b) $y = e^{2x}$, (c) $y = \cos 2x$, (d) $y = x^{-x/2}$

32 Rocket propulsion occurs as a result of the conservation of momentum, the burnt fuel being expelled in one direction with a momentum which is balanced by the momentum of the rocket in the other direction. The thrust T of a rocket is thus given by the rate of change of momentum, i.e.

$$T = -v\frac{dm}{dt}$$

where dm/dt is the rate of change of mass of the rocket as a result of fuel expulsion and v the fuel velocity. Derive an equation relating the rate of change of thrust with time.

33 For an object oscillating with simple harmonic motion, the displacement x varies with time t in the way described by the equation $x = A \sin \omega t$. Derive an equation for the acceleration.

2 Techniques of differentiation

2.1 Basic rules of differentiation Chapter 1 showed that differentiation from first principles of y, which is a function of x, with respect to x is largely a matter of finding limiting values as δx approaches zero and how by the use of that technique a table of commonly used derivatives can be obtained. This table enabled functions such as x^n and $\sin ax$ to be differentiated. But how do we tackle such functions as $2x + 4x^2$, $4 \sin 2x$, $x^2 \sin 2x$, $4\,e^{\sin 2x}$, etc.? This chapter is about the rules of differentiation and the techniques that can be used to enable such functions to be differentiated. Chapter 3 shows how such rules and techniques can be used in engineering problems.

The following list indicates the rules and techniques that are considered in this chapter:

1 Sums of functions (2.1.1).
2 Functions multiplied by constants (2.1.2).
3 Products of functions (2.1.3).
4 Quotients of functions (2.1.4).
5 Using the chain rule for functions of functions (2.2).
6 Implicit differentiation (2.3).
7 Inverse functions (2.4).
8 Parametric differentiation (2.5).
9 Logarithmic functions (2.6).

2.1.1 Differentiation of sums of functions

Consider the sum of a number of functions of a common variable, e.g. $y = x + x^2$. In general we can write such a sum as:

$$y = u + v - w$$

where u, v and w are all functions of x. The total increase in y due to an increase of δx in x is equal to the algebraic sum of all the increases in u, v and w. Thus

$$\delta y = \delta u + \delta v - \delta w$$

Hence we can write

$$\frac{\delta y}{\delta x} = \frac{\delta u}{\delta x} + \frac{\delta v}{\delta x} - \frac{\delta w}{\delta x}$$

When $\delta x \to 0$, we have

$$\frac{dy}{dx} = \frac{du}{dx} + \frac{dv}{dx} - \frac{dw}{dx} \qquad [1]$$

Similar relationships can be derived for the algebraic sum of any finite number of functions of x.

Example

Find dy/dx for the following:
(a) $y = x - x^2$, (b) $y = \sin x + \cos x$, (c) $y = 2 + e^{3x}$

(a) Item 2 in table 1.1 gives the derivatives for each of these terms. Thus, using the rule given by equation [1],

$$\frac{dy}{dx} = \frac{d}{dx}(x) - \frac{d}{dx}(x^2)$$

and so

$$\frac{dy}{dx} = 1 - 2x$$

(b) Items 5 and 6 in table 1.1 give the derivatives for each of these terms. Thus, using the rule given by equation [1],

$$\frac{dy}{dx} = \frac{d}{dx}(\sin x) + \frac{d}{dx}(\cos x)$$

and so

$$\frac{dy}{dx} = \cos x - \sin x$$

(c) Items 1 and 3 in table 1.1 give the derivatives for each of these terms. Thus, using the rule given by equation [1],

$$\frac{dy}{dx} = \frac{d}{dx}(2) + \frac{d}{dx}(e^{3x})$$

and so

$$\frac{dy}{dx} = 0 + 3\,e^{3x}$$

Review problems

1 Find dy/dx for the following:

   (a) $y = 2 + 3x + 4x^2$, (b) $y = x + \sin 4x$, (c) $y = x^2 - x^3$,

   (d) $y = e^{2x} - e^{3x}$, (e) $y = 2 - e^{-3x}$, (f) $y = \cos 2x - \sin 2x$

2 The e.m.f. E produced by a thermocouple is related to the temperature of the hot junction t by $E = at + bt^2$, where a and b are constants. Determine the rate of change of e.m.f. with temperature.

3 The current i in a circuit consisting of an inductance L in series with a resistance R varies with time t according to the equation

$$i = I(1 - e^{-Rt/L})$$

Determine the rate of change of current with time.

2.1.2 Differentiation of a function multiplied by a constant

Consider the differentiation of some function of x, with respect to x, when it is multiplied by some constant a, with a being a real number. Thus if we have $y = x^2$ then dy/dx $= 2x$. Now if we have $ay = ax^2$, what is d(ay)/dx?

   An increase in ay is a times the increase in y. Hence we can write

$$\delta(ay) = a\delta y$$

Thus, dividing both sides of the equation by δx,

$$\frac{\delta(ay)}{\delta x} = a\frac{\delta y}{\delta x}$$

Thus, in the limit as $\delta x \to 0$,

$$\frac{d}{dx}(ay) = a\frac{dy}{dx} \qquad\qquad [2]$$

Thus the differential coefficient of ay, where a is a constant, is a times the differential coefficient of y.

Example

Determine the value of dy/dx for $y = 3x^2$.

Using equation [2], then

$$\frac{d}{dx}(3x^2) = 3\frac{d}{dx}(x^2) = 6x$$

Review problems

4 Find dy/dx for the following:

(a) $y = 2x + 3x^4$, (b) $y = 5 \sin 2x$, (c) $y = 2x^5 + 3 \sin 2x$

5 The displacement x with time t of an object is given by

$$x = 10t + 20t^2$$

Determine the rate of change of x with time t, i.e. the velocity.

2.1.3 Differentiation of products of functions

Consider the problem of determining dy/dx when we have a product of functions, e.g. $y = x \sin x$. In general we can consider such a problem to be the differentiation of the product of u and v, both being functions of x, with respect to x.

$$y = uv$$

When x increases by δx then both u and v will increase. The product then becomes

$$y + \delta y = (u + \delta u)(v + \delta v)$$

$$= uv + u\,\delta v + v\,\delta u + \delta u\,\delta v$$

But $y = uv$, hence

$$\delta y = u\,\delta v + v\,\delta u + \delta u\,\delta v$$

and so, dividing throughout by δx,

$$\frac{\delta y}{\delta x} = u\frac{\delta v}{\delta x} + v\frac{\delta u}{\delta x} + \frac{\delta u}{\delta x}\delta v$$

In the limit when $\delta x \rightarrow 0$ then δv will also tend to 0. Thus

$$\frac{d}{dx}(uv) = u\frac{dv}{dx} + v\frac{du}{dx} \qquad\qquad [3]$$

Note that the derivative of a product is *not* the product of the derivatives.

This rule can be extended to enable the derivative to be found for the product of any, finite, number of functions of x. Thus, for example, for $y = uvw$, where u, v and w are all functions of x, we can consider this to be the product of two factors (uv) and w and use equation [2] to obtain

$$\frac{d}{dx}(uvw) = uv\frac{dw}{dx} + w\frac{d}{dx}(uv)$$

Then by applying equation [2] again to the product uv,

$$\frac{d}{dx}(uvw) = uv\frac{dw}{dx} + w\left(u\frac{dv}{dx} + v\frac{du}{dx}\right)$$

$$= uv\frac{dw}{dx} + uw\frac{dv}{dx} + vw\frac{du}{dx} \qquad [4]$$

The differential coefficient is thus obtained by multiplying the differential coefficient of each term in turn by all the other factors and adding the results.

This gives us a method of differentiating x^n, since x^n is merely $x \times x \times x \times x \times \dots$ for n terms. Thus

$$\frac{d}{dx}(x^n) = x^{n-1}\frac{d}{dx}x + x^{n-1}\frac{d}{dx}x + x^{n-1}\frac{d}{dx}x + \dots \; n \text{ times}$$

$$= nx^{n-1}$$

This is the result quoted in table 1.1 as item 2.

Example

Find dy/dx for the following:

(a) $y = x \sin x$, (b) $y = x^2 e^{2x}$, (c) $y = (1 + x^2)(2 + x^2)$,

(d) $y = (2 + 3x)^3$

(a) This is the product of x and $\sin x$. Thus, using equation [3],

$$\frac{d}{dx}(x \sin x) = x\frac{d}{dx}(\sin x) + \sin x\frac{d}{dx}(x) = x \cos x + \sin x$$

(b) This is the product of x^2 and e^{2x}. Thus, using equation [3],

$$\frac{d}{dx}(x^2 e^{2x}) = x^2\frac{d}{dx}(e^{2x}) + e^{2x}\frac{d}{dx}(x^2) = 2x^2 e^{2x} + 2x\, e^{2x}$$

(c) This is the product of $(1 + x^2)$ and $(2 + x^2)$. Thus, applying equation [3],

$$\frac{d}{dx}[(1+x^2)(2+x^2)] = (1+x^2)\frac{d}{dx}(2+x^2) + (2+x^2)\frac{d}{dx}(1+x^2)$$

$$= 2x(1+x^2) + 2x(2+x^2) = 6x + 4x^3$$

(d) This can be considered to be the product of three terms, each of which is $(2 + 3x)$. Thus, applying equation [4],

$$\frac{d}{dx}(2+3x)^3 = (2+3x)^2\frac{d}{dx}(2+3x) + (2+3x)^2\frac{d}{dx}(2+3x)$$

$$+ (2+3x)^2\frac{d}{dx}(2+3x)$$

$$= 9(2+3x)^2$$

Review problems

6　Find dy/dx for the following:

(a) $y = x^2\cos 4x$, (b) $y = x\,e^{-2x}$, (c) $y = \sin x \cos 2x$,

(d) $y = (2 + x^2)\sin x$, (e) $y = e^{2x}\cos 3x$, (f) $y = (2 + x^2)(1 + x^3)$,

(g) $y = (x + 2)^2$, (h) $y = (x^2 + x)^4$

7　A current i in a circuit is exponentially damped and given by the following equation. Determine an expression for the rate of change of current with time t.

$i = 2\,e^{-3t}\cos 100t$

8　The distance s in metres travelled by an object is related to the time t in seconds by the equation

$s = 2t^2(3t + 1)$

Determine the rate of change of distance with time at $t = 2$ s.

9　The distance s in metres travelled by an object is related to the time t in seconds by the equation

$s = \sqrt{t}\,(1 + t)$

Determine the rate of change of distance with time at $t = 4$ s.

2.1.4 Differentiation of quotients of functions

Consider the differentiation of y with respect to x, where $y = u/v$ with both u and v being functions of x. For example, we might be obtaining dy/dx for $y = x^2/(3x + 5)$. When x increases by δx then u and v will both increase, becoming $u + \delta u$ and $v + \delta v$. Thus

$$y + \delta y = \frac{u + \delta u}{v + \delta v}$$

Hence

$$\delta y = \frac{u + \delta u}{v + \delta v} - y = \frac{u + \delta u}{v + \delta v} - \frac{u}{v}$$

$$= \frac{uv + v\delta u - uv - u\delta v}{v(v + \delta v)} = \frac{v\delta u - u\delta v}{v(v + \delta v)}$$

Dividing by δx gives

$$\frac{\delta y}{\delta x} = \frac{v\dfrac{\delta u}{\delta x} - u\dfrac{\delta v}{\delta x}}{v(v + \delta v)}$$

In the limit when δx, and also δu and δv, tend to 0 we have

$$\frac{dy}{dx} = \frac{v\dfrac{du}{dx} - u\dfrac{dv}{dx}}{v^2} \qquad\qquad [5]$$

Example

Determine the derivative of the function

$$y = \frac{2x^2 - 3x}{x + 5}$$

Using the quotient rule given by equation [5], with $u = 2x^2 - 3x$ and $v = x + 5$, then

$$\frac{dy}{dx} = \frac{(x + 5)(4x - 3) - (2x^2 - 3x)}{(x + 5)^2}$$

$$= \frac{4x^2 + 17x - 15 - 2x^2 + 3x}{(x + 5)^2}$$

$$= \frac{2x^2 + 20x - 15}{(x + 5)^2}$$

Example

Determine the derivative of the function:

$$y = \frac{xe^x}{\cos x}$$

This example involves both a quotient and a product. Using the quotient rule we have

$$\frac{dy}{dx} = \frac{\cos x \frac{d}{dx}(xe^x) - xe^x(-\sin x)}{\cos^2 x}$$

We can differentiate the product xe^x using the product rule, and obtain $x\,e^x + e^x$. Hence

$$\frac{dy}{dx} = \frac{e^x(\cos x + x\cos x + x\sin x)}{\cos^2 x}$$

Review problems

10 Determine the derivatives of the following functions:

(a) $y = \dfrac{x^2}{2+3x}$, (b) $y = \dfrac{2x}{x^2+1}$, (c) $y = \dfrac{x^2+4}{x^2-4}$, (d) $y = \dfrac{x}{1+\sqrt{x}}$,

(e) $y = \dfrac{xe^{4x}}{\sin x}$, (f) $y = \dfrac{(x-1)^2}{x^2+1}$, (g) $y = \dfrac{(x-3)(x+2)}{x(x-1)}$

11 Use the quotient rule to determine the derivative of:
   (a) $y = \tan\theta$, taking $\tan\theta$ as equal to $\sin\theta/\cos\theta$,
   (b) $y = \cot\theta$,
   (c) $y = \sec\theta$

12 Determine the derivative of

$$y = \frac{x^2}{\tan x}$$

2.2 Differentiation of a function of a function

This procedure is concerned with dealing with situations where we have y as a function of u and u as a function of x, i.e. functions of functions. This is a situation that may be artificially produced so that we can differentiate a function such as $y = (3x + 2)^2$. We put $u = 3x + 2$. Then $y = u^2$. We now have y as a function of u and u a function of x. We can thus obtain dy/du and du/dx. What we require from this is dy/dx.

In general, when we have y as a continuous function of u and

u a continuous function of x, then when x increases to $x + \delta x$ then u increases to $u + \delta u$ and y to δy. We can thus write

$$\frac{\delta y}{\delta x} = \frac{\delta y}{\delta u} \times \frac{\delta u}{\delta x}$$

But dy/dx is the limiting value of $\delta y/\delta x$ as $\delta x \to 0$. Thus

$$\frac{dy}{dx} = \lim_{\delta x \to 0} \left[\frac{\delta y}{\delta u} \times \frac{\delta u}{\delta x} \right]$$

$$= \lim_{\delta x \to 0} \frac{\delta y}{\delta u} \lim_{\delta x \to 0} \frac{\delta u}{\delta x}$$

But as dx tends to 0 then, since u is a function of x, du also tends to 0. Thus we can write

$$\frac{dy}{dx} = \lim_{\delta u \to 0} \frac{\delta y}{\delta u} \lim_{\delta x \to 0} \frac{\delta u}{\delta x}$$

and so

$$\frac{dy}{dx} = \frac{dy}{du} \times \frac{du}{dx} \qquad [6]$$

This equation is often referred to as the *chain rule* and is used whenever we want to differentiate a composite group of functions. The rule can be extended to any number of intermediate functions, e.g.

$$\frac{dy}{dx} = \frac{dy}{du} \times \frac{du}{dv} \times \frac{dv}{dx} \qquad [7]$$

Consider the function $y = (3x + 2)^2$. To differentiate this we put $u = 3x + 2$. Then $y = u^2$ and hence $du/dx = 3$ and $dy/du = 2u$ Thus, using the chain rule (equation [6])

$$\frac{dy}{dx} = 2u \times 3 = 6(3x + 2)$$

Example

Determine the derivative of the function $y = (1 - x)^3$.

Let $u = 1 - x$. Then $y = u^3$ and hence $dy/du = 3u^2$ and $du/dx = -1$. Thus, using the chain rule,

$$\frac{dy}{dx} = 3u^2 \times (-1) = -3(1 - x)^2$$

Example

Determine the derivative of the function

$$y = \left(\frac{x^2}{x-1}\right)^3$$

Let $u = x^2/(x-1)$. Then $y = u^3$. The chain rule then gives

$$\frac{dy}{dx} = 3u^2 \times \frac{d}{dx}\left(\frac{x^2}{x-1}\right)$$

Using the procedure for determining the derivative of a quotient, i.e. equation [5], then

$$\frac{dy}{dx} = 3u^2 \times \frac{(x-1)2x - x^2}{(x-1)^2} = 3\left(\frac{x^2}{x-1}\right)^2 \times \frac{x^2 - 2x}{(x-1)^2}$$

$$= \frac{3x^5(x-2)}{(x-1)^4}$$

Review problems

13 Determine the derivatives of the following functions:

(a) $y = (x+2)^5$, (b) $y = \sin(2x+1)$, (c) $y = \sin^2 x$,

(d) $y = \sqrt{(1+x^3)}$, (e) $y = \sin\sqrt{(1+x^2)}$, (f) $y = 4\,e^{\sin x}$,

(g) $y = \left(\frac{x-2}{2x^2+1}\right)^3$, (h) $y = \frac{1}{(x^3 - 2x^2 - 1)^3}$,

(i) $y = \left(x + \frac{1}{x}\right)^3$, (j) $y = \frac{1}{1 - \cos 2x}$

2.2.1 Related rates

The chain rule can be used to find the rates of change of two or more variables that are related by changing with respect to some common variable, e.g. time. Thus suppose we have the function $y = x^2$ and we know the rate of change of x with time t, i.e. dx/dt, and are required to find the rate of change of y with time. Using the chain rule we can write

$$\frac{dy}{dt} = \frac{dy}{dx} \times \frac{dx}{dt}$$

Since $dy/dx = 2x$ then

$$\frac{dy}{dt} = 2x\frac{dx}{dt}$$

Example

Water is poured into a container in the form of an inverted right circular cone (figure 2.1) with a semi-vertical angle of 45° at the rate of 0.002 m³/s. Determine the rate at which the water surface rises when the depth of the water in the cone is 0.50 m.

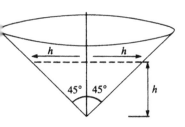

Fig. 2.1 Example

When the depth of the water is h then, because the semi-vertical angle is 45°, the radius of the water surface is also h. The volume V of the water at depth h is given by $V = \pi r^2 h/3 = \pi h^3/3$. Hence

$$\frac{dV}{dh} = \pi h^2$$

Using the chain rule

$$\frac{dV}{dt} = \frac{dV}{dh} \times \frac{dh}{dt}$$

Thus

$$0.002 = \pi h^2 \times \frac{dh}{dt}$$

$$\frac{dh}{dt} = \frac{0.002}{\pi h^2} = \frac{0.002}{\pi \times 0.50^2} = 2.54 \times 10^{-3} \text{ m/s}$$

Review problems

14 A gas expands at a constant temperature and the pressure p and volume V are related by $pV = $ a constant (Boyle's law). Initially the volume is $1\,m^3$ when the pressure is 10^5 Pa. The pressure is then decreased at the rate of 1 kPa/s. Determine the rate at which the gas expands when the volume is 1.20 m³.

15 Determine the rate at which the area of a circle increases when at a radius of 15 mm if the radius is increasing at the rate of 5 mm/min.

16 A ground camera is being used to film the lift-off of a rocket from its launch pad. The rocket rises vertically so that its vertical distance y, in metres, is related to time t, in seconds from launch, by $y = 10t^2$. Determine the rate of change of the angle of elevation of the camera 10 s after launch if the camera is situated horizontally 600 m from the launch pad.

2.3 Implicit differentiation

An equation involving two variables and of the form $y = f(x)$ is said to be in the *explicit* form. Thus, for example, we might have $y = 2x^2 + 3$. Such an equation is in the form $y = f(x)$. However, there are many functions which are not specified explicitly and are only implied by an equation. Thus if we have $y^2 = 2x + 3$ then y is only defined *implicitly* by the equation. Although y depends on x the equation is not in the form $y = f(x)$. One of the variables, y in this case, is not isolated. In such situations, to obtain the derivative dy/dx we use a method called *implicit differentiation*.

This involves differentiating both sides of the equation with respect to x. Thus for $y^2 = 2x + 3$ we have

$$\frac{d}{dx}(y^2) = \frac{d}{dx}(2x + 3)$$

To obtain $d(y^2)/dx$ we use the chain rule. Thus

$$\frac{d(y^2)}{dx} = \frac{d(y^2)}{dy} \times \frac{dy}{dx}$$

and so

$$2y\frac{dy}{dx} = 2$$

Thus

$$\frac{dy}{dx} = \frac{1}{y}$$

Example

Determine dy/dx given that $y^2 + 5y - x^2 = 3$.

Differentiating with respect to x gives

$$\frac{d}{dx}(y^2) + \frac{d}{dx}(5y) - \frac{d}{dx}(x^2) = \frac{d}{dx}(3)$$

Using the chain rule we can write

$$\frac{d(y^2)}{dx} = \frac{d(y^2)}{dy} \times \frac{dy}{dx}$$

and

$$\frac{d(5y)}{dx} = \frac{d(5y)}{dy} \times \frac{dy}{dx}$$

Hence

$$2y\frac{dy}{dx} + 5\frac{dy}{dx} - 2x = 0$$

and so

$$\frac{dy}{dx} = \frac{2x}{2y+5}$$

Review problems

17 Determine dy/dx given:

(a) $y^3 + y^2 - 3y - x^2 = 7$, (b) $3y^2 = 2x^2 + x$,

(c) $y^4 + xy^3 + x^2 = 2$, (d) $y^3 + 2x^2 - x = 1$,

(e) $y^2 - xy + x^2 = 2$, (f) $y^2 = (1 + 2x)^2$

2.4 Derivatives of inverse functions

If we have a function y which is a continuous function of x then the rate of change of y with x is dy/dx. However, if we have x as a continuous function of y then the rate of change of x with y is dx/dy. How are these rates of change related if the relationship between y and x is the same in both cases? For example, we might have $y = x^2$ and so obtain dy/dx = $2x$. The equation can, however, be written as $x = \sqrt{y}$ and so dx/dy = $\frac{1}{2}y^{-1/2}$.

We can use the chain rule to derive the relationship. Thus we must have

$$\frac{dy}{dx} \times \frac{dx}{dy} = 1$$

and so

$$\frac{dy}{dx} = \frac{1}{dx/dy} \qquad [8]$$

Thus for the example considered above where dy/dx = $2x$, then dx/dy = $1/2x = 1/(2y^{1/2})$.

Example

Determine dy/dx for the function described by $y^2 + 2y = x$.

It is a simpler operation to first obtain dx/dy by writing the equation as $x = y^2 + 2y$. Thus

$$\frac{dx}{dy} = 2y + 2$$

Hence, using equation [8],

$$\frac{dy}{dx} = \frac{1}{2y + 2}$$

Review problems

18 Determine dy/dx for the functions described by the following equations:

(a) $y^2 + 5y = x$, (b) $y^3 + 2y^2 = x$

2.4.1 Inverse trigonometric functions

If $y = x^2$ then the inverse function is $x = \sqrt{y}$. The inverse function of $y = \sin x$ is written as $x = \sin^{-1} y$ or $x = \arcsin y$. Note that $\sin^{-1} y$ does *not* mean $1/\sin x$. Thus, suppose we have $y = \sin \pi/4$, then the inverse is $\pi/4 = \sin^{-1} y$. This can be read as: $\pi/4$ radians is the angle whose sine is y. However, we do have a problem since $\pi/4$ rad or $45° = \sin^{-1} 0.707$, but we can also have other angles which equal $\sin^{-1} 0.707$, e.g. $3\pi/4$ or $135°$, $9\pi/4$ or $405°$, etc. There are an infinite number of values of x for a given value of y. Thus restrictions have to be placed on the inverse function if we are to have a one-to-one relationship for the value of an inverse. The restriction used is that $\sin^{-1} y$ is to lie between $-\pi/2$ and $+\pi/2$ (or $-90°$ and $+90°$), only then is a one-to-one relationship possible and a unique value occurs.

Consider the function $y = \sin^{-1}x$. Figure 2.2 shows the graph of this function and the restriction imposed on the function of it being restricted to the range $-\pi/2$ to $+\pi/2$. This function can be rewritten as $x = \sin y$. Therefore we have $dx/dy = \cos y$. Hence, using equation [8],

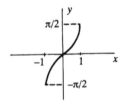

Fig. 2.2 $y = \sin^{-1}x$

$$\frac{dy}{dx} = \frac{1}{\cos y}$$

Given the relationship $y = \sin^{-1}x$ then we have described an angle y in a right-angled triangle which has a sine of x. The triangle is thus as shown in figure 2.3, the base being obtained by the use of the Pythagoras theorem. Thus $\cos y = \sqrt{(1 - x^2)}/1$ and so

Fig. 2.3 $y = \sin^{-1}x$

$$\frac{d}{dx}(\sin^{-1}x) = \frac{1}{\sqrt{1 - x^2}} \qquad [9]$$

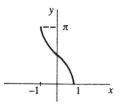

Fig. 2.4 $y = \cos^{-1}x$

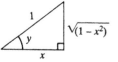

Fig. 2.5 $y = \cos^{-1}x$

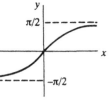

Fig. 2.6 $y = \tan^{-1}x$

Fig. 2.7 $y = \tan^{-1}x$

Consider the function $y = \cos^{-1}x$. Figure 2.4 shows its graph. For there to be a one-to-one relationship y is restricted to values between 0 and π. This function can be rewritten as $x = \cos y$ and so $dx/dy = -\sin y$. Hence, using equation [8],

$$\frac{dy}{dx} = -\frac{1}{\sin y}$$

Given the relationship $y = \cos^{-1}x$ then we have described an angle y in a right-angled triangle which has a cosine of x. The triangle is thus as shown in figure 2.5, the perpendicular being obtained by the use of the Pythagoras theorem. Thus $\sin y = \sqrt{(1 - x^2)}/1$ and so

$$\frac{d}{dx}(\cos^{-1}x) = -\frac{1}{\sqrt{1-x^2}} \qquad [10]$$

Consider the function $y = \tan^{-1}x$. Figure 2.6 shows the graph. In order to have a one-to-one relationship y is restricted to values between $-\pi/2$ and $+\pi/2$. This function can be written as $x = \tan y$. Therefore $dx/dy = \sec^2 y$. Hence, using equation [8],

$$\frac{dy}{dx} = \frac{1}{\sec^2 y} = \cos^2 y$$

Given the relationship $y = \tan^{-1}x$ then we have described an angle y in a right-angled triangle which has a tangent of x. The triangle is thus as shown in figure 2.7, the hypotenuse being obtained by the use of the Pythagoras theorem. Thus $\cos y = 1/\sqrt{(1 + x^2)}$ and so

$$\frac{d}{dx}(\tan^{-1}x) = \frac{1}{1+x^2} \qquad [11]$$

In a similar way, provided we impose restrictions on the inverses, we can obtain the derivatives of $\operatorname{cosec}^{-1}x$, $\sec^{-1}x$ and $\cot^{-1}x$.

$$\frac{d}{dx}(\operatorname{cosec}^{-1}x) = -\frac{1}{x\sqrt{x^2-1}} \qquad [12]$$

$$\frac{d}{dx}(\sec^{-1}x) = \frac{1}{x\sqrt{x^2-1}} \qquad [13]$$

$$\frac{d}{dx}(\cot^{-1}x) = -\frac{1}{1+x^2} \qquad [14]$$

The restricted ranges used to define the inverse trigonometric functions are given in table 2.1.

Table 2.1 Restricted ranges of inverse trigonometric functions

Function	Range of values of y
$y = \sin^{-1}x$	$-\pi/2 \le y \le \pi/2$
$y = \cos^{-1}x$	$0 \le y \le \pi$
$y = \tan^{-1}x$	$-\pi/2 < y < \pi/2$
$y = \text{cosec}^{-1}x$	$-\pi/2 \le y \le \pi/2, y \ne 0$
$y = \sec^{-1}x$	$0 \le y \le \pi, \ y \ne \pi/2$
$y = \cot^{-1}x$	$0 < y < \pi$

Example

Determine y for the equation $y = \sin^{-1}(-0.5)$.

This equation can be rewritten as $\sin y = -0.5$. Hence, within the restricted range $-\pi/2$ to $+\pi/2$, we have $y = -\pi/6$ or $-30°$.

Example

Determine x in the equation $\tan^{-1}(x + 2) = \pi/4$.

This equation can be rewritten as

$$x + 2 = \tan \pi/4 = 1$$

Hence $x = -1$.

Example

Determine dy/dx for the equation $y = \sin^{-1}3x$.

Let $u = 3x$. Then $y = \sin^{-1}u$. We then have $du/dx = 3$ and using equation [9]

$$\frac{dy}{du} = \frac{1}{\sqrt{1 - u^2}}$$

Then, using the chain rule,

$$\frac{dy}{dx} = \frac{dy}{du} \times \frac{du}{dx} = \frac{3}{\sqrt{1 - u^2}} = \frac{3}{\sqrt{1 - 9x^2}}$$

Example

Determine dy/dx for the equation $y = \sin^{-1}\sqrt{x}$.

If we let $u = \sqrt{x}$, then we have $y = \sin^{-1}u$. Equation [9] gives

$$\frac{dy}{du} = \frac{1}{\sqrt{1-u^2}}$$

and we also have

$$\frac{du}{dx} = \tfrac{1}{2}x^{-1/2}$$

Thus, using the chain rule,

$$\frac{dy}{dx} = \frac{dy}{du} \times \frac{du}{dx} = \tfrac{1}{2}x^{-1/2} \times \frac{1}{\sqrt{1-u^2}} = \frac{1}{2\sqrt{x}\sqrt{1-x}}$$

$$= \frac{1}{2\sqrt{x-x^2}}$$

Example

Determine dy/dx for the equation $y = x\sin^{-1}x/2$.

This is the derivative of a product. Thus

$$\frac{dy}{dx} = x\frac{d}{dx}\left(\sin^{-1}\frac{x}{2}\right) + \sin^{-1}\frac{x}{2}$$

If we let $u = x/2$ then $du/dx = 1/2$ and, using equation [9],

$$\frac{d}{du}(\sin^{-1}u) = \frac{1}{\sqrt{1-u^2}}$$

Thus, using the chain rule,

$$\frac{d}{dx}\left(\sin^{-1}\frac{x}{2}\right) = \frac{1}{2\sqrt{1-x^2/4}} = \frac{1}{\sqrt{4-x^2}}$$

Hence

$$\frac{dy}{dx} = \frac{x}{\sqrt{4-x^2}} + \sin^{-1}\frac{x}{2}$$

Review problems

19 Determine *y* for the following equations:

(a) $y = \tan^{-1}\sqrt{3}$, (b) $y = \cos^{-1}0$, (c) $y = \sin^{-1}\sqrt{3}/2$

20 Determine dy/dx for the following equations:

(a) $y = \tan^{-1}3x$, (b) $y = \cos^{-1}2x$, (c) $y = \sin^{-1}x/5$,

(d) $y = \tan^{-1}(1 + 3x)$, (e) $y = \sin^{-1}2x^2$, (f) $y = x^2\cos^{-1}(x - 1)$,

(g) $y = \sec^{-1}\sqrt{x}$, (h) $y = \cot^{-1}x^2$

2.4.2 Inverse hyperbolic functions

See section 1.4.5 for a discussion of hyperbolic functions and their derivatives. Consider the inverse of hyperbolic functions. Inverse hyperbolic functions are written in the form $\sinh^{-1}x$ or arsinh *x*. If we have the function $y = \sinh^{-1}x$ then we can write $x = \sinh y$ and so

$$\frac{dx}{dy} = \cosh y$$

But $\cosh y = \frac{1}{2}(e^y + e^{-y})$ and $\sin y = \frac{1}{2}(e^y - e^{-y})$ (see the section on hyperbolic functions in the Appendix). Thus

$$\cosh^2 y - \sinh^2 y = \tfrac{1}{4}(e^y + e^{-y})^2 - \tfrac{1}{4}(e^y - e^{-y})^2$$

$$= \tfrac{1}{4}(e^{2y} + e^{-2y} + 2) - \tfrac{1}{4}(e^{2y} + e^{-2y} - 2)$$

$$= 1$$

Thus we can write for dx/dy

$$\frac{dx}{dy} = \sqrt{1 + \sinh^2 y} = \sqrt{1 + x^2}$$

Thus

$$\frac{d}{dx}(\sinh^{-1}x) = \frac{1}{\sqrt{1 + x^2}} \qquad [15]$$

Consider the function $y = \cosh^{-1}x$. We can write $x = \cosh y$ and so (see section 1.4.5)

$$\frac{dx}{dy} = \sinh y$$

Thus, using $\cosh^2 y - \sinh^2 y = 1$,

$$\frac{dx}{dy} = \sqrt{\cosh^2 y - 1} = \sqrt{x^2 - 1}$$

Hence

$$\frac{d}{dx}(\cosh^{-1} x) = \frac{1}{\sqrt{x^2 - 1}} \qquad\qquad [16]$$

Consider the function $y = \tanh^{-1} x$. We can write $x = \tanh y$ and so (see section 1.4.5)

$$\frac{dx}{dy} = \operatorname{sech}^2 y$$

But $1 - \tanh^2 y = \operatorname{sech}^2 y$ (see the Appendix). Hence

$$\frac{dx}{dy} = 1 - \tanh^2 y = 1 - x^2$$

and so

$$\frac{d}{dx}(\tanh^{-1} x) = \frac{1}{1 - x^2} \qquad\qquad [17]$$

Example

Determine dy/dx for the equation $y = \sinh^{-1} x/3$.

Let $u = x/3$, then $y = \sinh^{-1} u$ and $du/dx = 1/3$. Thus, using equation [15],

$$\frac{dy}{du} = \frac{1}{\sqrt{1 + u^2}}$$

Hence, using the chain rule,

$$\frac{dy}{dx} = \frac{dy}{du} \times \frac{du}{dx} = \frac{1}{3\sqrt{1 + x^2/9}} = \frac{1}{\sqrt{9 + x^2}}$$

Review problems

21 Determine dy/dx for the following:

(a) $y = \cosh^{-1} 3x$, (b) $y = \sinh^{-1}(\sin 2x)$,

(c) $y = \cosh^{-1}\sqrt{(1 + x^2)}$, (d) $\tanh^{-1} 3x$

2.5 Parametric differentiation

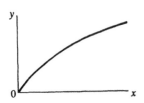

Fig. 2.8 A projectile

If an object is thrown vertically upwards then we can describe its motion by an equation involving the vertical displacement y and time t, i.e. $y = f(t)$. However, consider an object which is thrown in a direction at some angle to the horizontal, i.e. a projectile which has displacements in both the vertical and horizontal directions which are changing with time (figure 2.8). We thus have three variables, namely displacement in the vertical direction y, displacement in the horizontal direction x and time t, and two equations $y = f(t)$ and $x = g(t)$. The f and the g indicate that we can have two different functions of t. Thus we might, for example, have the equations $y = 10t - 4.9t^2$ and $x = 5t$. By eliminating t from the two equations we can arrive at a relationship between y and x, namely that $y = 2x - 0.196x^2$. Relationships between y and x which are specified in terms of some other variable, t in this case, are said to be *parametric equations* with this other variable being called the *parameter*.

There are many situations where the relationships between two variables y and x are most easily obtained by first determining the relationship between each of the variables and some third variable, a parameter. For example, with an electrical circuit we might know how the input current varies with time and how the output voltage varies with time. From these two parametric equations we could deduce how the output voltage depends on the input current.

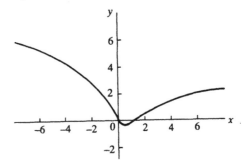

Fig. 2.9 $y = t^2 - t$ and $x = t^3$

Consider a graph of y against x which is defined by means of the parametric equations $y = t^2 - t$ and $x = t^3$. Figure 2.9 shows the graph. To determine the derivative dy/dx we can use the chain rule. Thus

$$\frac{dy}{dx} = \frac{dy}{dt} \times \frac{dt}{dx}$$

Thus, since $dy/dt = 2t - 1$ and $dx/dt = 3t^2$, then

$$\frac{dy}{dx} = (2t - 1) \times \frac{1}{3t^2} = \frac{2t - 1}{3t^2}$$

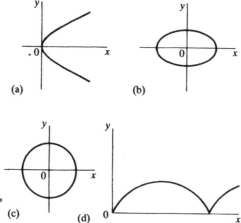

Fig. 2.10 (a) Parabola, (b) ellipse, (c) circle, (d) cycloid

The following are some commonly encountered curves (see figure 2.10) with their descriptions by parametric equations:

Parabola	$x = at^2$	$y = 2at$
Ellipse	$x = a \cos \theta$	$y = b \sin \theta$
Circle	$x = a \cos \theta$	$y = a \sin \theta$
Cycloid	$x = a(\theta - \sin \theta)$	$y = a(1 - \cos \theta)$

The example that follows illustrates, for one particular curve, how such parametric equations are obtained, the example being for the cycloid.

Example

Determine the parametric equations which can be used to describe the curve, a cycloid, traced out by a point P on the circumference of a circle of radius a as the circle rolls along a straight line.

Figure 2.11 shows the circle in its initial position, with P at the origin, and then some time t later when the angle through which the radius linking P to the circle centre has rotated an angle θ. The horizontal distance travelled $x = OD - BD$.

$$BD = AC = a \sin(180° - \theta) = a \sin \theta$$

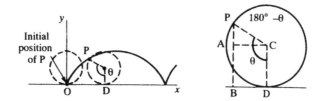

Fig. 2.11 Example

Since the circle rolls along the x-axis then OD must equal the arc PD. Thus

$$OD = a\theta$$

Thus the horizontal distance travelled

$$x = OD - BD = a\theta - a \sin \theta$$

The vertical distance travelled $y = BA + AP$.

$$AP = a \cos(180° - \theta) = -a \cos \theta$$

Thus the vertical distance travelled

$$y = BA + AP = a - a \cos \theta$$

Example

Determine the rate of change of y with respect to x for the cycloid in the previous example.

For the horizontal displacement

$$x = a\theta - a \sin \theta$$

and so

$$\frac{dx}{d\theta} = a - a \cos \theta$$

For the vertical displacement

$$y = a - a \cos \theta$$

and so

$$\frac{dy}{d\theta} = a \sin \theta$$

Thus, using the chain rule

$$\frac{dy}{dx} = \frac{dy}{d\theta} \times \frac{d\theta}{dx} = \frac{a \sin \theta}{a - a \cos \theta} = \frac{\sin \theta}{1 - \cos \theta}$$

Review problems

22 A function $y = f(x)$ is defined by the parametric equations $y = t^2$ and $x = t^3$. Determine dy/dx.

23 A function $y = f(x)$ is defined by the parametric equations $y = 2t(t + 1)$ and $x = 2t + 3$. Determine dy/dx.

24 An ellipse is defined by the parametric equations $x = 2 \cos \theta$ and $y = 3 \sin \theta$. Determine dy/dx and hence the gradient of the tangent at $\theta = \pi/3$.

25 A parabola is defined by the parametric equations $x = 2t^2$ and $y = 4t$. Determine dy/dx and hence the gradient of the tangent at $t = 2$.

2.6 Logarithmic functions

By taking logarithms of some functions prior to differentiation it is possible to simplify the process of differentiation. This is because the laws of logarithms change products into additions, quotients to subtractions and powers to products. With logarithms to any base

$$\log(A \times B) = \log A + \log B \qquad [18]$$

$$\log(A/B) = \log A - \log B \qquad [19]$$

$$\log A^n = n \log A \qquad [20]$$

A and B can be constants or functions of some variable. Generally with differentiation, logarithms to base e are used, these being denoted by ln.

Example

Determine dy/dx for the function $y = 2x \sin x$.

Taking logarithms to base e,

$$\ln y = \ln 2x + \ln(\sin x)$$

Differentiating with respect to x,

$$\frac{d}{dx}(\ln y) = \frac{d}{dx}(\ln 2x) + \frac{d}{dx}[\ln(\sin x)]$$

Using the chain rule we have for the first term

$$\frac{d(\ln y)}{dx} = \frac{d(\ln y)}{dy} \times \frac{dy}{dx} = \frac{1}{y}\frac{dy}{dx}$$

For the second term, if we let $u = 2x$ then the chain rule gives

$$\frac{d(\ln 2x)}{dx} = \frac{d(\ln u)}{du} \times \frac{du}{dx} = \frac{2}{u} = \frac{2}{2x}$$

For the third term, if we let $u = \sin x$ then the chain rule gives

$$\frac{d}{dx}[\ln(\sin x)] = \frac{d(\ln u)}{du} \times \frac{du}{dx} = \frac{1}{u} \times \cos x = \frac{\cos x}{\sin x}$$

Hence

$$\frac{1}{y}\frac{dy}{dx} = \frac{1}{x} + \frac{\cos x}{\sin x}$$

and so

$$\frac{dy}{dx} = y\left(\frac{1}{x} + \frac{\cos x}{\sin x}\right) = 2x \sin x \left(\frac{1}{x} + \frac{\cos x}{\sin x}\right)$$

$$= 2 \sin x + 2x \cos x$$

Example

Determine dy/dx for the function $y = e^{x^2-2}$.

Taking logarithms to base e,

$$\ln y = x^2 - 2$$

Differentiating with respect to x,

$$\frac{d}{dx}(\ln y) = \frac{d}{dx}(x^2) - \frac{d}{dx}(2)$$

Using the chain rule the first term is

$$\frac{d(\ln y)}{dx} = \frac{d(\ln y)}{dy} \times \frac{dy}{dx} = \frac{1}{y}\frac{dy}{dx}$$

Thus

$$\frac{1}{y}\frac{dy}{dx} = 2x$$

and so

$$\frac{dy}{dx} = 2xy = 2x e^{x^2-2}$$

Review problems

26 Determine dy/dx for the following functions:

(a) $y = x^2 \sin x$, (b) $y = 2^{3x+1}$, (c) $y = 2^x \tan x$, (d) $y = \dfrac{2 - x^2}{\tan x}$,

(e) $y = e^{2x} \sin^3 x \cos^2 x$, (f) $y = e^{-2x} x^3 \ln x$.

Further problems

27 Find dy/dx for the following:

(a) $y = 3x + \sin 2x$, (b) $y = 2 + x^2 + 3x^3 + x^4$,

(c) $y = 2 \sin x + \cos 3x$, (d) $y = x^2 \cos 4x$,

(e) $y = (2 + 3x) \sin x$, (f) $y = 2 e^x \sin 2x$,

(g) $y = (2 + e^{-x}) \sin 4x$, (h) $y = (1 + 2x)(2 + x^3)$, (i) $y = (x + 2)^3$,

(j) $y = (x^2 + 2x)^4$, (k) $y = x \cos x$, (l) $y = (5x - 1) \sin x$,

(m) $y = \dfrac{x^2 - 1}{\cos x}$, (n) $y = \dfrac{2x + 3}{3x + 2}$, (o) $y = \dfrac{5x}{x^2 + 3}$, (p) $y = \cot 2x$,

(q) $y = \dfrac{x \sin x}{(x - 1)}$, (r) $y = (2x^3 - 3x)^4$, (s) $y = \cos^5 x$,

(t) $y = 2 e^{2x+1}$, (u) $y = \sin(2x + 3)$, (v) $y = \dfrac{1}{(1 + 3x)^7}$,

(w) $y^2 + 2x^2 = 3$, (x) $2y^3 - 3x^2 + 4x = 1$, (y) $y^3 + 3xy + 2x^2 = 4$,

(z) $y = \sin^{-1} 5x$, (aa) $y = \cos^{-1} 4x$, (ab) $y = \tan^{-1} 2x$,

(ac) $y = \tan^{-1}(1 - x)$, (ad) $y = \cos^{-1} \sqrt{1 - x}$, (ae) $y = x \sin^{-1} x^2$,

(af) $y = \dfrac{\sin^{-1} x}{x}$, (ag) $y = \sin^{-1}(2 \sin x)$, (ah) $y = \cot^{-1} \sqrt{x^2 - 1}$,

(ai) $y = \cosh^{-1} 4x$, (aj) $y = \operatorname{sech}^{-1}(2x - 1)$, (ak) $y = x \sinh^{-1} x$,

(al) $y = \dfrac{(x + 2)(x + 1)}{(x + 3)^2}$, (am) $y = 2x^{1/2} \sin 2x$, (an) $y = e^{2x} \tan x$

28 In a chemical reaction the amount A of a substance produced
is related to the time t by

$A = 20t - 4t^2$

Determine how the rate at which the substance is produced is
related to the time.

29 The velocity v of an object is related to time t by

$$v = - 100 \sin 5t$$

Determine how the acceleration, i.e. the rate of change of velocity with time, is related to the time.

30 The frequency f of vibration of a string is related to the tension T in the string by

$$f = 100 \sqrt{T}$$

Hence determine the rate of change of frequency with tension.

31 The length L of a metal rod is a function of temperature t and is given by

$$L = L_0(1 + at + bt^2)$$

Determine the rate at which length is increasing with respect to temperature when $t = 100$.

32 A spherical balloon is inflated by pumping air into it at the rate of 10 m³/min. Determine the rate at which the radius increases when the radius is 4 m.

33 The Doppler effect gives, for the frequency f_0 heard by an observer when the source of a sound of frequency f_s is moving away from the observer with a velocity v,

$$f_0 = \frac{f_s}{1 + v/c}$$

where c is the velocity of sound. Determine the rate of change of the observed frequency with respect to the velocity v.

34 The radius of a sphere is increasing at the rate of 50 mm/min. Determine the rate of change of the volume when the radius is 150 mm.

35 A hot air balloon rises vertically from the ground with a constant velocity of 1 m/s. An observer is situated horizontally 30 m from the balloon launch point. What will be the rate of change of the elevation of the balloon from the observer when the balloon is 30 m above the ground?

36 The displacement x of an object is related to the time t by the equation $x^2 - 3t^2 = 9$. Determine an equation for the velocity.

37 Show for $y = \sec^{-1}x$ that

$$\frac{dy}{dx} = \frac{1}{x\sqrt{x^2 - 1}}$$

38 Show for $y = \text{sech}^{-1}x$ that

$$\frac{dy}{dx} = -\frac{1}{x\sqrt{1-x^2}}$$

39 A function $y = f(x)$ is defined by the parametric equations $y = 5t(t + 2)$ and $x = 2t$. Determine dy/dx.

40 An ellipse is defined by the parametric equations $x = 5 \cos \theta$ and $y = 2 \sin \theta$. Determine dy/dx and hence the gradient of the tangent at $\theta = \pi/6$.

41 A parabola is defined by the parametric equations $x = 5t^2$ and $y = 10t$. Determine dy/dx and hence the gradient of the tangent at $t = 3$.

42 A circle of radius 1 cm rolls around the outside of a circle of radius 2 cm without slipping. Using the angle θ shown in figure 2.12 as the parameter, determine a pair of parametric equations which can be used to define the curve traced by a point P on the circumference of the smaller circle. (Note: the curve is called an epicycloid.)

43 The function $y = f(x)$ is defined by the parametric equations $x = 1 - t$ and $y = t^2$. Determine the point on the graph of y against x at which the tangent is horizontal, i.e. $dy/dx = 0$.

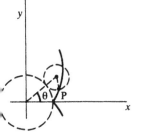

Fig. 2.12 Problem 42

3 Applications of differentiation

3.1 Mechanics applications

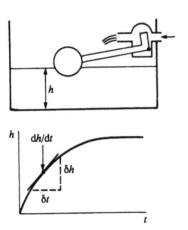

Fig. 3.1 Float-controlled valve system

This chapter aims to give some idea of the types of application in engineering that can occur for differentiation. As indicated in chapter 1, dy/dx is a measure of the rate at which y increases with respect to x. If we consider a graph of y plotted against x then dy/dx is the gradient of the graph at a point. For example, we might have a graph showing how the height of liquid in a container varies with time as a consequence of liquid entering through a float-controlled valve (figure 3.1). If the change in height is δh in a time δt then $\delta h/\delta t$ is the average rate of change of height over the time interval δt. In the limit as we make dt smaller and smaller so that it tends to 0, then we end up with what can be considered the rate of change of height with time dh/dt at an instant of time.

There are many situations in engineering involving rates of change. Thus, in mechanics, velocity is the rate of change of displacement with time and acceleration is the rate of change of velocity with time.

3.1.1 Velocity and acceleration

The *velocity* of a particle is the rate at which its displacement from a fixed point changes with time (figure 3.2). Thus if the displacement changes by δx in a time δt then the average velocity over that time interval is $\delta x/\delta t$. As $\delta t \to 0$ then $\delta x/\delta t$ tends to a limiting value dx/dt which is the velocity v at some instant of time t.

$$\lim_{\delta t \to 0} \frac{\delta x}{\delta t} = \frac{dx}{dt}$$

and so

$$v = \frac{dx}{dt} \qquad [1]$$

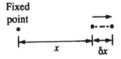

Fixed point

Fig. 3.2 Change in displacement

dx/dt will be positive if x increases as t increases and negative if x decreases as t increases.

The *acceleration* of a particle is the rate at which the velocity changes. Thus if the velocity changes by δv in a time δt then the average acceleration over that time interval is $\delta v/\delta t$. As $\delta t \to 0$ then $\delta v/\delta t$ tends to a limiting value dv/dt which is the acceleration a at some instant of time t.

$$\lim_{\delta t \to 0} \frac{\delta v}{\delta t} = \frac{dv}{dt}$$

and so

$$a = \frac{dv}{dt} \qquad [2]$$

dv/dt will be positive if v increases as t increases and negative if v decreases as t increases. The term *retardation* is used for a negative acceleration.

We can use equation [1] to rewrite this as

$$a = \frac{d}{dt}\left(\frac{dx}{dt}\right) = \frac{d^2x}{dt^2} \qquad [3]$$

Acceleration can also be written in another form since equation [1] can be written, using the chain rule, as

$$a = \frac{dv}{dt} = \frac{dv}{dx} \times \frac{dx}{dt} = \frac{dv}{dx} \times v \qquad [4]$$

Example

The displacement x of a particle from some fixed point when oscillating with simple harmonic motion is given by the equation $x = A \cos \omega t$. Derive equations for the velocity v and the acceleration a.

Using equation [1],

$$v = \frac{d}{dt}(A \cos \omega t) = -A\omega \sin \omega t$$

Using equation [2],

$$a = \frac{d}{dt}(-A\omega \sin \omega t) = -A\omega^2 \cos \omega t$$

Since $x = A \cos \omega t$ then we can write $a = -\omega^2 x$. The acceleration therefore is proportional to the displacement from the fixed point and directed towards it, i.e. the particle suffers a retardation as x increases.

Example

A particle moves in a straight line so that its distance x, in metres, from a fixed point on the line varies with time t, in seconds, according to the equation

$$x = 9t - t^3$$

Hence determine the velocity and acceleration after 1 s.

Using equation [1],

$$v = \frac{\mathrm{d}}{\mathrm{d}t}(9t - t^3) = 9 - 3t^2$$

Thus after 1 s the velocity is 6 m/s. Using equation [2],

$$a = \frac{\mathrm{d}}{\mathrm{d}t}(9 - 3t) = -6t$$

Thus after 1 s the acceleration is –6 m/s². The minus sign shows that it is actually a retardation.

Example

The velocity of a particle varies inversely as the displacement from a fixed point. How does the acceleration vary with the displacement?

The relationship between the velocity v and the displacement x is of the form $v = k/x$, where k is a constant. Thus, differentiating v with respect to x, we have $\mathrm{d}v/\mathrm{d}x = -k/x^2$ Hence, using equation [4],

$$a = v\frac{\mathrm{d}v}{\mathrm{d}x} = -v\frac{k}{x^2} = -\frac{k}{x} \times \frac{k}{x^2} = -\frac{k^2}{x^3}$$

Thus the acceleration is inversely proportional to the cube of the displacement.

Review problems

1 The distance travelled by a particle in a straight line from a fixed point varies as the square root of the time elapsed since it left the fixed point. Derive relationships for the velocity and acceleration in terms of the time.

2 Determine the velocity and the acceleration after 2 s for a body for which the displacement x, in metres, is related to the time t, in seconds, by $x = 15 + 20t - 2t^2$.

3 Determine the initial velocity and acceleration of a particle for

which the displacement x, in metres, is related to the time t, in seconds, by $x = 2 + 3t^2 + t^4$.

4 Determine the velocity and acceleration of a body after a time t if the displacement x in metres is related to the time in seconds by the equation $x = 2 \sin 2t + 4 \cos 3t$.

5 A ball rolls from rest down an inclined plane. If the distance travelled down the plane x, in metres, is related to time t, in seconds, by $x = 3t^2$ determine, (a) how the velocity varies with time, (b) the velocity after the ball has rolled 4 m.

6 A ball, after being given an initial push, rolls down an inclined plane. If the distance travelled down the plane x, in metres, is related to time t, in seconds, by $x = 5t + 3t^2$ determine (a) how the velocity varies with time, (b) the velocity after the ball has rolled 4 m.

3.1.2 Angular velocity and acceleration

Consider an arm rotating in a circular path, as shown in figure 3.3. The *angular displacement* of the arm is the angle swept out by the rotation, the angle being in units of radians with one complete rotation of 360° being 2π radians. The *angular velocity* of the arm is the rate at which its angular displacement from a fixed position changes with time. Thus if the displacement changes by $\delta\theta$ in a time δt then the average angular velocity over that time interval is $\delta\theta/\delta t$. As $\delta t \to 0$ then $\delta\theta/\delta t$ tends to a limiting value $d\theta/dt$ which is the angular velocity ω at some instant of time t.

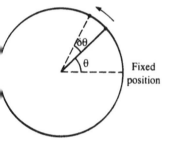

Fig. 3.3 Change in angular displacement

$$\omega = \lim_{\delta t \to 0} \frac{\delta\theta}{\delta t} = \frac{d\theta}{dt} \qquad [5]$$

$d\theta/dt$ will be positive if θ increases as t increases and negative if θ decreases as t increases.

The *angular acceleration* of the rotating arm is the rate at which the angular velocity changes. Thus if the angular velocity changes by $\delta\omega$ in a time δt then the average acceleration over that time interval is $\delta\omega/\delta t$. As $\delta t \to 0$ then $\delta\omega/\delta t$ tends to a limiting value $d\omega/dt$ which is the acceleration α at some instant of time t.

$$\alpha = \lim_{\delta t \to 0} \frac{\delta\omega}{\delta t} = \frac{d\omega}{dt} \qquad [6]$$

$d\omega/dt$ will be positive if ω increases as t increases and negative if ω decreases as t increases.

We can use equation [5] to rewrite equation [6] in the form

$$\alpha = \frac{d\omega}{dt} = \frac{d}{dt}\left(\frac{d\theta}{dt}\right) = \frac{d^2\theta}{dt^2} \qquad [7]$$

The angular acceleration can also be written in another form. Equation [6] can be written, using the chain rule, as

$$\alpha = \frac{d\omega}{dt} = \frac{d\omega}{d\theta} \times \frac{d\theta}{dt} = \frac{d\omega}{d\theta} \times \omega \qquad [8]$$

Example

A flywheel rotates such that the angle θ, in radians, rotated is related to the time t, in seconds, by $\theta = 10 + 15t - t^2$. Determine (a) an equation describing how the angular velocity changes with time, (b) an equation describing how the angular acceleration varies with time, (c) the time taken for the flywheel to come to rest.

(a) Using equation [5], then

$$\omega = \frac{d\theta}{dt} = 15 - 2t$$

(b) Using equation [6], then

$$\alpha = \frac{d\omega}{dt} = -2$$

The angular acceleration is thus a constant retardation of 2 rad/s^2.
(c) The angular velocity will be zero when the flywheel is at rest. Thus we must have

$$15 - 2t = 0$$

Hence the time taken to come to rest is 7.5 s.

Example

Figure 3.4 shows a rod which is maintained in contact with a cam by a spring. The rod is constrained so that it can only be moved in the horizontal direction. Determine equations for the velocity and the acceleration of the rod, in terms of the angular velocity and the angular acceleration of the cam, when the cam rotates.

Consider the triangle ABC. For such a triangle BC/AC = cos θ. Thus, since BC = $x/2$ and AC = r,

$$x = 2r \cos \theta$$

Differentiating this with respect to time t, then by the chain rule (see section 2.1.4)

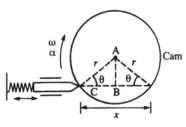

Fig. 3.4 Example

$$\frac{dx}{dt} = -2r \sin \theta \frac{d\theta}{dt}$$

But $dx/dt = v$, the velocity of the rod, and $d\theta/dt = \omega$, the angular velocity of the cam. Thus

$$v = -2r\omega \sin \theta$$

We can obtain the acceleration a of the rod by differentiating the above equation. Thus

$$a = \frac{dv}{dt} = -2r\frac{d\omega}{dt} \sin \theta - 2r\omega \cos \theta \frac{d\theta}{dt}$$

and so

$$a = -2r\alpha \sin \theta - 2r\omega^2 \cos \theta$$

Review problems

7 A wheel rotates in such a way that the angle θ through which it rotates is related to the time t by the equation

$$\theta = 10 + 10t - t^2$$

Determine how the angular velocity and the angular acceleration vary with time.

8 Determine the time a rotating wheel will take to come to rest if the angle θ through which it rotates is related to the time t, in seconds, by

$$\theta = 20 + 4t - 2t^2$$

9 Figure 3.5 shows a mechanism, involving a crank AB and connecting rod BC, for the linear motion back and forth of the piston C in a cylinder when the crank rotates. Derive equations for the angular velocity and angular acceleration of the connecting rod in terms of the angular velocity and angular acceleration of the crank. Assume ϕ is small enough for $\cos \phi \approx 1$.

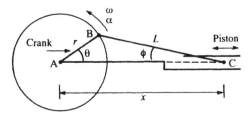

Fig. 3.5 Problem 9

3.1.3 Motion along curves

One way which can be used to specify motion along curved paths is to specify both the x and y coordinates of the moving point as functions of time t. This set of equations is referred to as a set of parametric equations (see section 2.5). An example of an object moving in a curved path is a projectile, with one of the parametric equations describing how its vertical height varies with time, and the other equation how its horizontal displacement varies with time.

Example

An object moves in a curved path specified by the parametric equations $x = 3t$ and $y = 12t - 5t^2$, x and y being in metres with t in seconds. Determine the magnitude and direction of the velocity when $t = 2$ s.

These equations describe an object following a parabolic path and could, for example, describe the path of a projectile. Figure 3.6 shows such a path. Differentiating each equation gives for the velocity components in the x and y directions

$$v_x = \frac{dx}{dt} = 3$$

$$v_y = \frac{dy}{dt} = 12 - 10t$$

Thus at $t = 2$ s then $v_x = 3$ m/s and $v_y = -8$ m/s. The minus sign indicates that the velocity is in a downward direction. The velocity resulting from these two components is (see figure 3.6)

$$v = \sqrt{3^2 + (-8)^2} = 8.5 \text{ m/s}$$

This velocity is at an angle to the horizontal of

$$\theta = \tan^{-1}\left(\frac{-8}{3}\right) = -69.4°$$

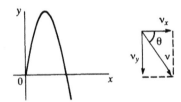

Fig. 3.6 Example

Example

A particle moves along a curve described by the equation $y = x^2$ in such a way that the velocity component in the x direction is always 2 m/s. Determine the magnitude and direction of the velocity when the particle has the coordinates (2, 4) m.

Differentiating the equation by t (see section 2.3) gives

$$\frac{dy}{dt} = 2x\frac{dx}{dt}$$

Thus $v_y = 2xv_x$. Hence with $v_x = 2$ m/s and $x = 2$ m, then $v_y = 8$ m/s and so the resultant velocity is $\sqrt{(2^2 + 8^2)} = 8.24$ m/s at an angle to the horizontal of $\tan^{-1}(8/2) = 76.0°$.

Review problems

10 An object moves along a curved path with its coordinates being given by $x = 4t$ and $y = 10t - 5t^2$, with x and y being in metres and t in seconds. What is the magnitude and direction of the velocity when $t = 0.5$ s?

11 A particle moves in a parabolic path described by the equation $y^2 = 12x$ in such a way that the velocity component in the x direction is constant at 10 m/s. Determine the resultant velocity when the particle is at a point with the coordinates $(3, 6)$.

12 A particle moves along a curved path with its coordinates being given by $x = 1 + 2 \cos t$ and $y = 3 + 2 \sin t$. Show that the particle has a resultant velocity which is constant at all times.

3.1.4 Force and acceleration

When a resultant *force* F acts on an object of mass m then it accelerates with an acceleration a given by

$$F = ma$$

Thus we can write

$$F = m\frac{dv}{dt} = m\frac{d^2x}{dt^2} \qquad [9]$$

The *work* W done by a force is the product of the force and the distance moved in the direction of the force. Thus, if x is this direction, then $W = Fx$. *Power* P is the rate of doing work. Thus

$$P = \frac{dW}{dt} = \frac{d}{dt}(Fx)$$

If the force is constant, then

$$P = Fv \qquad [10]$$

Example

A particle of mass 0.1 kg moves in a straight line so that its distance x, in metres, from a fixed point varies with time t, in seconds, according to the equation $x = 3 + 2t^2$. Determine the force acting on the particle after 1 s.

The velocity is given by differentiating the equation for the displacement. The acceleration is given by a second differentiation. Thus

$$v = \frac{dx}{dt} = 4t$$

$$a = \frac{d^2x}{dt^2} = 4$$

The acceleration is constant with time. The force is given by

$$F = ma = 0.1 \times 4 = 0.4 \text{ N}$$

Review problems

13 A particle moves in a straight line so that its distance x from a fixed point is given by the equation $x = 2 + 3t + 4t^2$. Prove that the force causing the motion is constant.
14 Derive an equation showing how the force acting on a particle of mass m changes with time when it is moving in a straight line with a displacement x which is related to the time t by the equation $x = 2 + 3 \sin t$.

3.2 Electrical applications

Electrical *current i* is the rate of movement of charge q with time t, i.e.

$$i = \frac{dq}{dt} \qquad [11]$$

A current of 1 ampere (A) is a movement of charge at the rate of 1 coulomb (C) per second.

For charge to move through a circuit element, energy must be supplied. The energy required to move a charge of 1 C through a circuit element is called the *potential difference* or voltage across that element. This is the energy that is dissipated in the charge moving through the element. Thus if δw is the energy required for an element of charge δq then the voltage v is

$$v = \frac{\delta w}{\delta q}$$

In the limit as $\delta q \to 0$ then

$$v = \frac{dw}{dq} \qquad\qquad [12]$$

The voltage is in volts (V) when the energy w is in joules (J) and the charge q in coulombs (C).

When there is a current, i.e. charge is being moved through a circuit, then energy has to be continually expended. The rate at which energy is expended is called the *power*. Power p can thus be expressed as

$$p = \frac{dw}{dt} \qquad\qquad [13]$$

The power is in watts (W) when the energy is in joules (J) and the time in seconds (s). We can write equation [13], using the chain rule, in the form

$$p = \frac{dw}{dq} \times \frac{dq}{dt}$$

Thus, using equations [11] and [12],

$$p = vi \qquad\qquad [14]$$

Power is thus the product of the potential difference across an element and the current through it.

Example

Derive an equation for the current through a circuit element if the charge q, in coulombs, entering the element is related to the time t, in seconds, by $q = 10t$.

Using equation [11],

$$i = \frac{dq}{dt} = \frac{d}{dt}(10t) = 10 \text{ A}$$

Review problems

15 Derive an equation for the current through a circuit element if the charge q, in coulombs, entering the element is related to the time t, in seconds, by $q = 2 \, e^{-3t}$.

16 What is the power dissipated in a circuit element if there is a potential difference of 2 V across it when charge is being moved through it at the rate of 4 C/s?

3.2.1 Capacitors and inductors

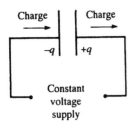

Charge →   Charge →

$-q$ $+q$

Constant
voltage
supply

Fig. 3.7 Capacitor being
charged

A capacitor is essentially just a pair of parallel conducting plates
separated by an insulator. When a capacitor is connected to a
constant voltage supply then one of the plates becomes negatively
charged and the other positively charged as a result of charge
flowing on to one of the plates and leaving the other plate (figure
3.7). The amount of charge gained by one plate has the same
magnitude as that lost by the other plate. If this charge is q then it
is found that q is directly proportional to the potential difference v
between the plates at the instant of time concerned. Hence $q \propto v$
and so

$$q = Cv$$

where C, the constant of proportionality, is termed the *capacitance*. Capacitance has the unit of a farad (F) when the voltage is
in volts (V) and the charge in coulombs (C). Using equation [11]
we can write the above equation as

$$i = \frac{dq}{dt} = C\frac{dv}{dt} \qquad [15]$$

An inductor can be considered to be essentially just a coil or
wire. When the current through the coil changes then the magnetic
flux generated by the current changes. This changes the magnetic
flux linked by the coil and hence an e.m.f. is induced. This induced
e.m.f. is in such a direction as to oppose the change producing it.
Thus if it was produced by an increasing current the direction of
the e.m.f. will be such as to slow down the increase. The induced
e.m.f. is proportional to the rate of change of linked flux. Hence,
since the amount of linked flux is proportional to the current we
have the induced e.m.f. proportional to the rate of change of
current. We can thus write

$$\text{e.m.f.} \propto -\frac{di}{dt}$$

The minus sign is because the e.m.f. opposes the change in current
responsible for it. This equation can then be written as

$$\text{e.m.f.} = -L\frac{di}{dt}$$

L, the constant of proportionality, is called the *inductance*. The
inductance has the unit of a henry (H) when the e.m.f. is in volts
(V), the current in amperes (A) and the time in seconds (s). If we
are concerned with a pure inductance, i.e. one which has only
inductance and no resistance, then there can be no potential

difference across the inductor due to resistance. To maintain the current through the inductor then the source must supply a potential difference v across the inductor which is just sufficient to cancel out the induced e.m.f. Thus the potential difference v across the inductor is

$$v = L\frac{di}{dt} \qquad\qquad [16]$$

Example

The voltage v, in volts, across a capacitor is continuously adjusted so that it varies with time t, in seconds, according to the equation $v = 2t$. Derive an equation indicating how the current varies with time for a capacitance of 2 mF.

Using equation [15],

$$i = C\frac{dv}{dt} = 0.002\frac{d}{dt}(2t) = 0.004 \text{ A}$$

The current is thus maintained constant at 4 mA.

Example

The current i, in amperes, through an inductor of inductance 2 H increases linearly with time t in seconds, being related to the time by the equation $i = 3t$. Hence derive a relationship for the voltage across the inductor.

Using equation [16],

$$v = L\frac{di}{dt} = 2\frac{d}{dt}(3t) = 6 \text{ V}$$

The voltage is thus constant at 6 V.

Review problems

17 The voltage v, in volts, across a capacitor varies with time t, in seconds, according to the equation

$$v = 4 \sin 2t$$

How will the current vary with time if the capacitance is 2 mF?

18 The voltage v, in volts, across a capacitor varies with time t, in seconds, according to the equation

$$v = 2 \sin(4t + 45°)$$

How will the current vary with time if the capacitance is 2 mF?

19 The current i, in amperes, through an inductor of inductance 50 mH varies with time t, in seconds, and is described by the equation

$$i = 10(1 - e^{-100t})$$

Determine how the potential difference across the inductor varies with time.

20 The current i, in amperes, through an inductor of inductance 100 mH varies with time t, in seconds, and is described by the equation

$$i = 10t\,e^{-2t}$$

Determine how the potential difference across the inductor varies with time.

21 The current i in a series RL circuit, when there is a step voltage input of V, varies with time t according to the equation

$$i = \frac{V}{R}(1 - e^{-Rt/L})$$

Derive an equation indicating how the voltage across the inductor varies with time.

3.2.2 Electric field strength

Forces act on electric charges when in electric fields. The *electric field strength E* can be defined as being the potential gradient, i.e.

$$E = -\frac{dV}{dx}$$

The minus sign is because the direction of the electric field is in the opposite direction to that in which the potential is increasing.

Consider an isolated positive charge $+q$ in an electric field and it being moved from a point at potential V to one, a distance δx away at potential $(V - \delta V)$, as illustrated in figure 3.8. The potential gradient is $\delta V/\delta x$. There is an electric field strength E of $-\delta V/\delta x$. The charge will experience a force F as a consequence of being in an electric field. The work done w in moving the charge the distance δx is thus

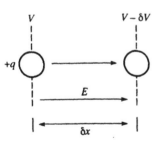

Fig. 3.8 Moving a charge in an electric field

$$w = F\,\delta x$$

Hence, since the potential difference between two points is the work done per unit charge moved between them, then

$$-\delta V = \frac{w}{q} = \frac{F\,\delta x}{q}$$

Hence

$$E = -\frac{dV}{dx} = \frac{F}{q} \qquad\qquad [17]$$

Thus the electric field strength is the force per unit charge.

Example

The potential V a distance r from the centre of a positive charge q is given by

$$V = -\frac{Q}{4\pi\varepsilon\varepsilon_0 r}$$

Derive an equation for the electric field strength at a distance r from the centre of the charge.

Using equation [17],

$$E = -\frac{dV}{dr} = \frac{Q}{4\pi\varepsilon\varepsilon_0 r^2}$$

Review problems

22 The variation of the potential V with distance x in the barrier layer between p- and n-type semiconductors is given by

$$V = \frac{A}{1 + e^{-kx}}$$

where A and k are constants. Derive an equation for the force acting on an electron with a charge $-e$.

3.3 Stationary points

Consider a graph of y against x, where y is a function of x, as in figure 3.9. The gradient of the graph is given by dy/dx. If dy/dx is greater than 0 then y is increasing as x increases, i.e. the gradient has a positive value. If dy/dx is less than 0, i.e. has negative values, then y is decreasing as x increases. However, if $dy/dx = 0$ then y is neither increasing or decreasing. Points where this occur are called

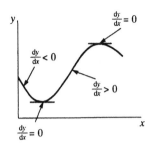

Fig. 3.9 Gradients

stationary points or *critical points* or *turning points*. At such points we must have the tangent to the graph parallel to the x-axis.

Example

Determine the point on a graph of y plotted against x at which the gradient is zero for the function $y = 2x^2 + 3x + 1$.

Differentiating the function gives

$$\frac{dy}{dx} = 4x + 3$$

Thus dy/dx = 0 when 4x + 3 = 0, i.e. x = -3/4. The value of y at this point is $y = 2(-3/4)^2 + 3(-3/4) + 1 = -1/8$. Thus the point has the coordinates (-3/4, -1/8).

Example

Determine the point on a graph of y plotted against x at which the gradient is zero for the function $y = x^3 + 2x + 5$.

Differentiating the function gives

$$\frac{dy}{dx} = 3x^2 + 2$$

Thus dy/dx = 0 when $3x^2 + 2 = 0$, i.e. $x^2 = -2/3$. There is no real solution, thus there is no real value of x at which the gradient is zero.

Review problems

23 Determine, for the following functions, the points for which dy/dx = 0:

(a) $y = 2x^3$, (b) $y = 4x^2 + 2x$, (c) $y = \sin 2x$, (d) $y = 1/x$

24 When a stone is thrown vertically upwards with an initial velocity u, the distance y moved vertically is related to the time t by the equation

$$y = ut - \tfrac{1}{2}gt^2$$

where g is the acceleration due to gravity. Determine the equation for the greatest height reached. (Hint: the greatest height is when the velocity becomes zero.)

3.3.1 Maxima and minima

There are many situations in engineering where we are concerned with establishing the minimum or maximum condition. Thus with an electrical circuit we might need to establish the condition for which maximum power is generated in a load resistance. With a beam which is bent we might need to determine the maximum bending moment. In all these situations we have effectively a graph of one function against another and need to establish the point on the graph at which it is a minimum or maximum. Such points occur when the gradient of the graph is zero.

Consider y to be a function of x, i.e. $y = f(x)$. Points on a graph of y plotted against x at which $dy/dx = 0$ can be:

1   a local maximum
2   a local minimum
3   a point of inflexion

Figure 3.10 illustrates these points. In figure 3.10(a), $dy/dx = 0$ at point P. This point is a local maximum because for all points in the vicinity the value of the function is greatest at this point. In figure 3.10(b), $dy/dx = 0$ at point P. This point is a local minimum because for all points in the vicinity the value of the function is smallest at this point. In the two graphs given in figure 3.10(c), $dy/dx = 0$ at point P. However, in neither of the graphs is there a local maximum or minimum. Such points as these are called *points of inflexion*.

If we consider the gradients in the vicinity of maxima, minima and points of inflexion, then:

1   at a maximum the gradient changes from positive to negative
2   at a minimum the gradient changes from negative to positive
3   at a point of inflexion the sign of the gradient does not change

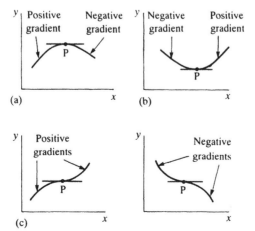

Fig. 3.10 (a) Local maximum, (b) local minimum, (c) point of inflexion

Since d^2y/dx^2 measures the rate of change of dy/dx with x then when a gradient changes from positive to negative it is decreasing and so d^2y/dx^2 must be negative. Hence at a maximum d^2y/dx^2 is negative. When a gradient changes from negative to positive then it is increasing and so d^2y/dx^2 is positive. Hence at a minimum d^2y/dx^2 is positive.

Example

Determine, and identify the form of, the turning points on a graph of the equation $y = x^3 - 9x^2 + 15x$.

Differentiating the equation gives

$$\frac{dy}{dx} = 3x^2 - 18x + 15 = 3(x - 1)(x - 5)$$

Thus when $dy/dx = 0$ then $x = 1$ or $x = 5$. Substituting these x values in the original equation gives the y values at these turning points. Thus the turning points are $(1, 7)$ and $(5, -25)$.

Consider the turning point at $(1, 7)$ and how the gradient changes from just prior to the point to just after it. If x is slightly less than 1 then $(x - 1)$ is negative and $(x - 5)$ is negative; dy/dx is positive. When x is slightly greater than 1 then $(x - 1)$ is positive and $(x - 5)$ is negative; dy/dx is negative. The gradient changes from positive to negative and so the point $(1, 7)$ is a maximum.

Consider the turning point at $(5, -25)$ and how the gradient changes from just prior to the point to just after it. If x is slightly less than 5 then $(x - 1)$ is positive and $(x - 5)$ is negative; dy/dx is negative. When x is slightly greater than 5 then $(x - 1)$ is positive and $(x - 5)$ is positive. Thus the gradient changes from negative to positive and so the point $(5, -25)$ is a minimum.

As an alternative to considering the gradients prior to and after the points we could consider d^2y/dx^2. Differentiating dy/dx gives

$$\frac{d^2y}{dx^2} = 6x - 18$$

At the point $(1, 7)$ then $d^2y/dx^2 = -12$, the condition for a maximum. At the point $(5, -25)$ then $d^2y/dx^2 = 12$, the condition for a minimum.

Figure 3.11 shows the graph of the function with the maximum and minimum identified.

Example

The displacement y in metres of an object is related to the time t in seconds by the equation $y = 1 + 4t - t^2$. Determine the maximum displacement.

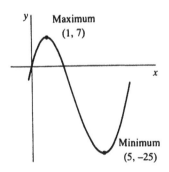

Fig. 3.11 Example

Differentiating the equation gives

$$\frac{dy}{dt} = 4 - 2t$$

dy/dt is 0 when $t = 2$ s. There is thus a turning point at the displacement $x = 1 + 8 - 4 = 5$ m. Since $d^2y/dx^2 = -2$ then the point is a maximum.

Example

If the sum of two numbers is 40, determine the values which will give the minimum value for the sum of their squares.

Let the two numbers be x and y. Then $x + y = 40$. We have to find the minimum value of S where

$$S = x^2 + y^2$$

We can rewrite this equation as

$$S = x^2 + (40 - x)^2 = x^2 + 1600 - 80x + x^2$$

$$= 2x^2 - 80x + 1600$$

Differentiating this equation gives

$$\frac{dS}{dx} = 4x - 80$$

This is a turning point when $x = 20$. We can check that this is a minimum by obtaining $d^2S/dx^2 = 4$. Since this is positive then we have a minimum. The two numbers are thus 20 and 20.

Example

Determine the dimensions of the largest rectangle that can be cut from a semicircular sheet with a radius r.

Figure 3.12 shows the situations. The area $A = 2xy$. Using the Pythagoras theorem then $r^2 = y^2 + x^2$. Hence we can write

$$A = 2x \sqrt{r^2 - x^2}$$

Differentiating, using the product rule, gives

$$\frac{dA}{dx} = 2x \frac{d}{dx}\left(\sqrt{r^2 - x^2}\right) + 2\sqrt{r^2 - x^2}$$

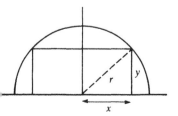

Fig. 3.12 Example

Let $u = r^2 - x^2$. Then $\sqrt{(r^2 - x^2)} = \sqrt{u}$. Thus, using the chain rule,

$$\frac{d}{dx}\left(\sqrt{r^2 - x^2}\right) = \frac{d}{du}(\sqrt{u}) \times \frac{du}{dx}$$

$$= \tfrac{1}{2}u^{-1/2} \times (-2x) = -\frac{x}{\sqrt{r^2 - x^2}}$$

Thus dA/dx is a maximum when

$$-\frac{2x^2}{\sqrt{r^2 - x^2}} + 2\sqrt{r^2 - x^2} = 0$$

i.e. when

$$-2x^2 + 2(r^2 - x^2) = 0$$

and $x = r/\sqrt{2}$. Then $y = \sqrt{(r^2 - x^2)} = \sqrt{(r^2 - r^2/2)} = r/\sqrt{2}$. The rectangle thus has the dimensions $2r/\sqrt{2} = r\sqrt{2}$ and $r/\sqrt{2}$.

Review problems

25 Determine, and identify the form of, the turning points on graphs of the following equations:

(a) $y = 2x^3 - 3x^2 - 12x + 12$, (b) $y = x - 2 \sin x$,

(c) $y = \dfrac{x^2 - 4x + 9}{x^2 + 4x + 9}$, (d) $y = (3x^2 + 1)^2$

26 The rate r at which a chemical reaction proceeds is related to the quantity x of chemical present by

$$r = k(a - x)(b + x)$$

Determine the maximum rate.

27 The bending moment M of a uniform beam of length L at a distance x from one end is given by

$$M = \tfrac{1}{2}wLx - \tfrac{1}{2}wx^2$$

where w is the weight per unit length of beam. Determine the point at which the bending moment is a maximum.

28 An electrical circuit consists of a voltage source of V and internal resistance r in series with a load of resistance R. The power P generated by the load is given by

$$P = \frac{V^2 R}{R + r}$$

Determine the value of load resistance at which the power dissipation will be a maximum and the value of this maximum power.

29 Determine the two numbers whose sum is 40 and for which the product is a maximum.

30 Determine the height and radius required for a cylinder with a volume 16π m^3 if it is to have the least surface area.

31 Determine the area of the largest rectangular piece of ground that can be enclosed by a perimeter fence of length 100 m.

32 Determine the dimensions an open box can have for maximum volume if constructed with a square base from a sheet of metal of area 108 cm^2.

33 The cost C of producing x units of an item is given by

$$C = 500 + 0.02x + 0.0001x^2$$

Determine the number that should be produced if the average cost per item, i.e. C/x, is to be a minimum.

34 An electrical voltage v is given by

$$v = 40 \cos 1000t + 15 \sin 1000t$$

Determine the maximum value of the voltage.

35 A circuit consists of two resistances R_1 and R_2 in parallel. The sum of the two resistances is 100 Ω. Determine the values of these resistances if, for a constant current entering the circuit, the power is to be a maximum.

3.4 Small changes

In chapter 1 the derivative dy/dx was defined as the limiting value of $\delta y/\delta x$ as δx tended to a zero value. Chapter 1 and then chapter 2 were concerned with how dy/dx could be obtained for functions. Frequently in engineering we are concerned with establishing what change in one quantity will be produced by a change in another. We can consider this to mean the determination of δy when there is a change in x of δx. In such situations it is common to assume that $\delta y/\delta x$ is the same as dy/dx and thus the derivation of dy/dx can be used to obtain $\delta y/\delta x$.

Example

The pressure p and volume v of an ideal gas are related by Boyle's law, namely $pv = $ a constant. What will be the percentage change in the volume of the gas when the pressure changes by 1%?

Differentiating the equation $p = k/v$, where k is a constant, gives

$$\frac{dp}{dv} = -\frac{k}{v^2} = -\frac{pv}{v^2}$$

Thus, considering small changes in the pressure and volume, we can write

$$\frac{\delta p}{p} = -\frac{\delta v}{v}$$

A fractional change in the pressure of 1/100 will thus result in a volume change of −1/100. The minus sign indicates that the volume is reduced by this fraction when the pressure increases. Thus there is a 1% reduction in the volume.

Review problems

36 The electrical resistance R of a metal changes with temperature θ according to the equation

$$R = R_0(1 + a\theta + b\theta^2)$$

where R_0, a and b are constants. Determine an equation for the increase in resistance that will occur for a small change in temperature of $\delta\theta$.

37 The periodic time T of oscillation of a simple pendulum of length L is given by

$$T = 2\pi\sqrt{\frac{L}{g}}$$

where g is the acceleration due to gravity. Determine the percentage change in the value of the periodic time if the length is increased by 1%.

3.4.1 Errors

There are invariably errors associated with measurements, the *error* of a measurement being the difference between the measured value and what is estimated to be the true value of it. As a consequence, the results of measurements are frequently quoted in terms of the measured value plus or minus the estimated error associated with that measurement. Thus if the true value is x and there is an error of δx, then the measured value is likely to be quoted as $x + \delta x$. In calculations of errors the approximation is generally made of δx being equal to dx.

Example

The diameter of a spherical ball is measured as 10.0 ± 0.2 mm. Estimate the maximum error in the volume of the ball calculated from this measurement.

The volume V of a sphere is given by $V = \frac{4}{3}\pi r^3$ and since $d = 2r$ by $V = \frac{1}{6}\pi d^3$. Differentiating this equation with respect to diameter d gives

$$\frac{\mathrm{d}V}{\mathrm{d}d} = \frac{1}{2}\pi d^2$$

Thus we write

$$\delta V = \frac{1}{2}\pi d^2 \, \delta d$$

The estimated error δd is ± 0.2 mm. Thus

$$\delta V = \frac{1}{2}\pi 10.0^2 \times (\pm 0.2) \text{ mm}^3$$

The maximum error is thus 31.4 mm^3.

Review problems

38 The results of measurements of the sides of a cube are quoted as 45 ± 1 mm. Estimate the maximum error in the volume calculated from this measurement.

39 The periodic time T of a simple pendulum of length L is given by

$$T = 2\pi \sqrt{\frac{L}{g}}$$

Measurements of the periodic time are being made to obtain a value for the acceleration due to gravity g. What will be the percentage error in the value of g if the periodic time is measured with an accuracy of 0.5%?

Further problems

40 Determine the velocity and acceleration after a time t for a particle which has a displacement x from a fixed point which changes with time t according to $x = 20 \cos 2t$.

41 Determine the initial velocity and acceleration for a particle which has a displacement x, in metres, from a fixed point which varies with time t, in seconds, according to the equation $x = 4 + 6t - 2t^2$.

42 A particle has a displacement x, in metres, which varies with time t, in seconds, according to $x = 12 + 30t - t^3$. At what time will the velocity be zero?

43 The velocity of a particle is proportional to the square of its displacement from a fixed point. How is the acceleration related to the displacement?

44 The velocity of a particle is inversely proportional to the square root of its displacement from a fixed point. How is the acceleration related to the velocity?

45 An object slides along a surface with its displacement x, in metres, from a fixed point being related to the time t, in seconds, by $x = 12t - 2t^2$. Determine (a) how the velocity will vary with time and (b) the time taken before the object comes to rest.

46 A flywheel rotates such that its angular displacement θ, in radians, is related to the time t, in seconds, by

$$\theta = 20 + 5t^2$$

What is the angular velocity after a time of 2 s?

47 A wheel rotates with an angular velocity ω which is related to the time t by

$$\omega = 10 - 2t^2$$

How does the angular acceleration vary with time?

48 Figure 3.13 shows a rigid rod AB. The end A is constrained to move only in the vertical y direction and the end B is so constrained as to move only in the horizontal x direction. If the end B moves in the x direction, determine equations relating the velocity v and the acceleration a of B with the angular velocity ω and angular acceleration α of the rod.

49 Derive an equation for the current through a circuit element if the charge q, in coulombs, entering the element is related to the time t, in seconds, by $q = 0.5\,e^{-2t}$.

50 The voltage v across a capacitor varies with time t. Derive equations for how the current varies with time if (a) $v = at$, with a being a constant, (b) $v = V \cos \omega t$.

51 The current i, in amperes, through an inductance of 1 mH varies with time t, in seconds, according to the equation $i = 10\,e^{-3t}$. Determine how the potential difference across the inductor varies with time.

52 The current i, in amperes, through an inductance of 10 mH varies with time t, in seconds, according to the equation

$$i = 4e^{-200t} - 4e^{-600t}$$

Determine how the potential difference across the inductor varies with time.

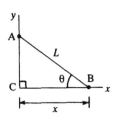

Fig. 3.13 Problem 48

53 Determine for the following functions the points on graphs of y against x at which the gradients are zero:

(a) $y = 2x^2 - x$, (b) $y = 2x^4$, (c) $y = 2 \cos 4x$, (d) $y = x^{0.1}$

54 Determine, and identify the form of, the turning points on graphs of the following equations:

(a) $y = 3 \sin x + 4 \cos x$, (b) $y = x^4 - 6x^2 + 8x + 10$,

(c) $y = x^2 + \dfrac{16}{x^2}$, (d) $y = 2x - e^x$, (e) $y = 3x^2(2x - 1)$

55 The e.m.f. E produced by a thermocouple is related to the temperature θ by

$$E = a\theta + b\theta^2$$

Determine the temperature at which the e.m.f. is a maximum.

56 The speed v, in metres per second, of a car when in first gear is related to the time t, in seconds, by

$$v = 4 + 3t - t^2$$

Determine the maximum speed of the car.

57 A particle moving in a curved path has the coordinates given by $x = 2 \cos 10t$ and $y = 2 \sin 10t$. Show that the path followed is circular with a constant velocity.

58 A particle moves in a parabolic path $y^2 = 4x$ so that it always has a velocity component parallel to the x-axis, this component being constant at 2 m/s. Determine the velocity along the curve when $x = 1$ m.

59 A rocket moves vertically upwards with its distance x from the ground related to the time t by

$$x = 50t - 5t^2$$

Determine (a) how the velocity varies with time, (b) the greatest height reached.

60 An alternating current i, in milliamperes, in a circuit varies with time t according to the equation

$$i = 20 \sin 1000t$$

What is the maximum rate of change of the current and at what times does it occur?

61 A rectangle has an area of 100 m^2. Determine the length of the sides for which the diagonal has minimum length.

62 Determine the maximum area a rectangle can have if inscribed inside a triangle of area A.

63 Light, travelling with a velocity v, is reflected from a flat surface. Show that the condition for the path taken by the light to occupy the minimum time is that the angle of incidence equals the angle of reflection.

64 Light travelling with a velocity v_1 in one medium is refracted on passing into a medium where the velocity becomes v_2. Show that the condition for the path taken by the light to occupy the minimum time is that $v_1/v_2 = \sin i/\sin r$, where i is the angle of incidence and r the angle of refraction.

65 The cost C of producing x units of a product is given by a setting up cost of £400 which applies regardless of how many items are produced and a cost which depends on the number of units produced and is given, in £, by $0.02x + 0.0002x^2$. Determine the production level which will minimise the average cost per item produced.

66 The deflection y of a propped cantilever beam at a distance x from the fixed end when carrying a uniform load of w per unit length is given by

$$y = \frac{1}{EI}\left(\frac{5wLx^3}{48} - \frac{wL^2x^2}{16} - \frac{wx^4}{24}\right)$$

with E and I being constants. Determine the position of maximum deflection.

67 Power cables have to be connected from a power station to a factory. The power station is on the bank on one side of a river of width 20 m and the factory is on the other bank which is a distance 70 m downstream. If it costs three times as much to take the cables underwater as it does to take them across land, what should be the path of the cable for minimum cost?

68 The velocity v of waves in deep water is given by the equation

$$v = \sqrt{\frac{g\lambda}{2\pi} + \frac{2\pi T}{\rho\lambda}}$$

where λ is the wavelength of the waves, T the surface tension of the water and ρ its density. Determine the wavelength which gives the minimum velocity.

69 The rate of loss of heat from a hot body is proportional to the surface area of that body. For a cylindrical hot water tank, what is the relationship between the height of the cylinder and its diameter which will result in the minimum loss of heat?

70 What is the percentage change in the periodic time of a simple pendulum if its length increases, as a result of thermal expansion, by 1%? The periodic time T is related to the length L of the pendulum by

$$T = 2\pi\sqrt{\frac{L}{g}}$$

where g is the acceleration due to gravity.

71 The diameter of a sphere can be measured with an accuracy of ±1 mm. What will be the percentage error in (a) the volume, (b) the surface area when calculated using the measured value?

$\mathbf{4}$ Numerical differentiation

4.1 Derivatives from discrete values

If a function is specified by means of a table of discrete values, rather than an algebraic expression, then the analytical methods described in earlier chapters cannot be used to determine derivatives. Such situations can occur when experimental measurements are used to obtain data, e.g. distance values at specific times for a moving object, and the function relating the data values is not known. In addition, the calculation of derivative values by means of computers tends to be done numerically, even when the function is known, by operating on discrete values obtained for the function. This chapter describes how derivatives can be obtained in such situations.

There are two ways of considering the methods. One is in terms of considering lines joining data values to represent chords between points on a graph of the function. The other, which generates the same equations, involves representing the function by a series. We can represent a function by a polynomial series, the data values then being points which the series must represent. In considering, in this chapter, various degees of sophistication of numerical methods the equations are developed by both these methods.

4.1.1 Taylor series

The basis used in this chapter for the development of numerical methods based on representing a function by a series, is in terms of the Taylor series. The following is a brief discussion of the series.

A function can be represented, provided enough terms are considered, by a polynomial series.

$$f(x) = A + Bx + Cx^2 + Dx^3 + Ex^4 + \dots \qquad [1]$$

where A, B, C, D, E, etc. are constants. Consider the value of this function at $x = a$. Then

$$f(a) = A + Ba + Ca^2 + Da^3 + Ea^4 + ... \qquad [2]$$

Equation [1] minus equation [2] enables us to eliminate constant A to give

$$f(x) - f(a) = B(x - a) + C(x^2 - a^2) + D(x^3 - a^3)$$
$$+ E(x^4 - a^4) + ... \qquad [3]$$

Differentiating equation [2] with respect to a gives

$$f'(a) = B + 2Ca + 3Da^2 + 4Ea^3 + ... \qquad [4]$$

where $f'(a)$ is used to represent the $d\{f(x)\}/da$. Multiplying this equation by $(x - a)$ and subtracting it from equation [3] eliminates B to give

$$f(x) - f(a) - (x - a)f'(a)$$

$$= C(x^2 - a^2) - 2Ca(x - a) + D(x^3 - a^3) - 3Da^2(x - a)$$
$$+ E(x^4 - a^4) - 4Ea^3(x - a) + ...$$

$$= C(x - a)^2 + D(x^3 - 3a^2x + 2a^3)$$
$$+ E(x^4 - 4a^3x + 3a^4) + ... \qquad [5]$$

Differentiating equation [4] with respect to a gives

$$f''(a) = 2C + 6Da + 12Ea^2 + ... \qquad [6]$$

where $f''(a)$ represents $d^2\{f(x)\}/da^2$. Multiplying this equation by $(x - a)^2/2$ and subtracting it from equation [5] eliminates C and gives

$$f(x) - f(a) - (x - a)f'(a) - \frac{(x - a)^2}{2}f''(a)$$

$$= D(x^3 - 3a^2x + 2a^3) - 3Da(x - a)^2$$
$$+ E(x^4 - 4a^3x + 3a^4) - 6Ea^2(x - a)^2 + ...$$

$$= D(x - a)^3 + E(x^4 - 6a^2x^2 + 8a^3x - 3a^4) + ...$$

In general we can write

$$f(x) = f(a) + \frac{x - a}{1!}f'(a) + \frac{(x - a)^2}{2!}f''(a)$$

$$+ \frac{(x - a)^3}{3!}f'''(a) + ... + \frac{(x - a)^n}{n!}f^{(n)}(a) \qquad [7]$$

where we have $f'''(a)$ representing $d^3\{f(x)\}/da^3$, $f^{iv}(a)$ representing $d^4\{f(x)\}/da^4$, etc. The equation is known as *Taylor's series* or *theorem*.

Example

Determine the polynomial series which gives an approximate value for the function $y = f(x)$, if at $x = 2$ we have $y = 1$, $dy/dx = 3$, $d^2y/dx^2 = 4$, and $d^3y/dx^3 = -6$.

Using Taylor's theorem, i.e. equation [7], then

$$y = f(x) = 1 + (x-2) \times 3 + \frac{(x-2)^2}{2} \times 4 + \frac{(x-2)^3}{6} \times (-6)$$

$$= 1 + 3(x-2) + 2(x-2)^2 - (x-2)^3$$

Review problems

1 Determine the polynomial series which gives an approximate value for the function $y = f(x)$ if at $x = 1$ we have $y = 0$, $dy/dx = 2$, $d^2y/dx^2 = 0$, and $d^3y/dx^3 = 6$.

2 Determine the polynomial series which gives an approximate value for the function $y = f(x)$ if at $x = 2$ we have $y = 3$, $dy/dx = 3$, $d^2y/dx^2 = 2$, and $d^3y/dx^3 = -12$.

4.2 Two-point equation

Consider the situation if we have just two data points, e.g. for a distance–time relationship, the distance after 1 s and the distance after 2 s. Since we have no knowledge of the function relating the distance and time, all we can do is make an assumption. The simplest assumption is that there is a straight line relationship.

The derivative is defined in terms of the gradient of a chord joining two points on a graph of the function, i.e. equation [2] chapter 1,

$$\frac{dy}{dx} = \lim_{\delta x \to 0} \frac{f(x + \delta x) - f(x)}{\delta x} \qquad [8]$$

Thus for a graph of y plotted against x, as in figure 4.1, then $f(x)$ is the value of y at $x = a$, i.e. point P, and $f(x + \delta x)$ is the value at $x = a + h$, i.e. point B; $[f(a + h) - f(a)]/\delta a$ is the gradient of the chord between P and B. When we have just two data points then we can take this gradient of the chord, i.e. the gradient of the straight line joining the two data points, to be an approximation of the derivative of the function at point P. Thus

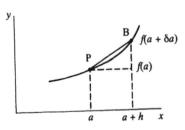

Fig. 4.1 Two-point equation

$$\frac{dy}{dx} \approx \frac{f(a+h) - f(a)}{h} \qquad\qquad [9]$$

This is termed a *two-point equation* since the derivative is estimated from just two points.

Example

The following data were obtained from measurement of the distance x travelled by a moving body as a function of time t. Estimate, using a two-point equation, the value of the derivative, i.e. the velocity, at a time of 1 s.

x in m	5.12	5.30
t in s	1.0	1.1

The value of the time interval is 0.1 s. Hence, using equation [9],

$$\frac{dx}{dt} \approx \frac{5.30 - 5.12}{0.1}$$

The estimate of the derivative is thus 1.8 m/s.

Example

The following are the results of measurements of distances x of a moving object at different times t.

x in m	5.00	6.05	7.20	8.45
t in s	1.0	1.1	1.2	1.3

Use the two-point equation to determine the derivative at a time of 1.0 s with (a) $h = 0.3$ s, (b) $h = 0.2$ s, (c) $h = 0.1$ s. Comment on the accuracy of the results if the equation relating the distance and time is $x = 5t^2$.

(a) With $h = 0.3$ s, then equation [9] gives

$$\frac{dx}{dt} \approx \frac{8.45 - 5.00}{0.3} \approx 11.5 \text{ m/s}$$

(b) With $h = 0.2$, then equation [9] gives

$$\frac{dx}{dt} \approx \frac{7.20 - 5.00}{0.2} = 11.0 \text{ m/s}$$

(c) With $h = 0.1$ s, then equation [9] gives

$$\frac{dx}{dt} \approx \frac{6.05 - 5.00}{0.1} = 10.5 \text{ m/s}$$

The exact solution for the derivative is obtained by differentiating $x = 5t^2$ to give $dx/dt = 10t = 10$ m/s. Thus the smaller the time interval between the data points the less the error in the derivative obtained by the two-point equation.

$h = 0.3$ m   error = 1.5 m/s

$h = 0.2$ m   error = 1.0 m/s

$h = 0.1$ m   error = 0.5 m/s

The error is proportional to h. The results quoted in (a), (b) and (c) are the average velocities over the time intervals concerned.

Review problems

3 Use the two-point equation to estimate of the derivative at $x = 2$ for the function giving the following data points:

y	4.25	4.61
x	2.0	2.1

4 The following data are distance x and time values t for a falling object. Using the two-point equation, estimate the velocity of the object at the times 1.0 s, 1.2 s and 1.4 s.

x in m	4.91	7.06	9.60	12.54
t in s	1.0	1.2	1.4	1.6

(Note: what is being calculated is the average velocity over the time intervals 1.0 s to 1.2 s, 1.2 s to 1.4 s and 1.4 s to 1.6 s.)

4.2.1 Taylor's theorem

In general we can describe any function by means of the *Taylor theorem* (see section 4.1.1), i.e. the value of the function $y = f(x)$ at $x = a$ is given by

$$f(x) = f(a) + \frac{x-a}{1!}f'(a) + \frac{(x-a)^2}{2!}f''(a)$$
$$+ \frac{(x-a)^3}{3!}f'''(a) + \dots + \frac{(x-a)^n}{n!}f^{(n)}(a) \qquad [10]$$

where $f'(a)$ is the value of dy/dx at $x = a$, $f''(a)$ is the value of d^2y/dx^2 at $x = a$, etc. The value of the function at $x = a + h$ will then be, when substituting $a + h$ for x in equation [10],

$$f(a + h) = f(a) + hf'(a) + \frac{h^2}{2}f''(a) + \ldots$$

Thus, rearranging this equation gives

$$f'(a) = \frac{f(a + h) - f(a)}{h} - \frac{h}{2}f''(a) - \ldots$$

The first part of this equation is the two-point equation [9]. We can thus obtain the two-point equation by considering just the first two terms in the series. The error in considering only these terms is due to the terms neglected, i.e. the $\frac{1}{2}hf''(a)$ and further terms and thus can be considered to be proportional to h. Hence, the smaller the interval between points the more accurate the estimate for the derivative. Halving h halves the possible error.

4.3 Three-point equation

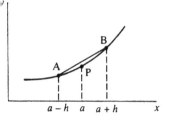

Fig. 4.2 Three-point equation

Consider the graph shown in figure 4.2. The derivative at point P, where $x = a$, can be obtained by drawing a chord symmetrically disposed about the point. With such a chord AB we have A with $x = a - h$ and B with $x = a + h$. The gradient of the chord AB is thus

$$\text{gradient} = \frac{f(a + h) - f(a - h)}{2h} \qquad [11]$$

This gradient is assumed to be an approximation to the tangent at point P and hence the required derivative. This equation is termed a *three-point equation* since it involves three data points, namely $f(a + h)$, $f(a)$ and $f(a - h)$.

Example

Use the three-point equation to arrive at a value for the derivative at $x = 3.0$ for the following data:

y	8.41	9.00	9.61
x	2.9	3.0	3.1

Taking h to be 0.1 and the ends of the chord to be at $x = 2.9$ and $x = 3.1$, then equation [11] gives

$$\text{gradient of chord} = \frac{9.61 - 8.41}{2 \times 0.1} = 6.0$$

This is now taken to be the estimate of the derivative at $x = 3.0$. (Note that the function is actually described by $y = x^2$ and so we have $dy/dx = 2x = 2 \times 3.0 = 6.0$.)

Review problems

5 Use the three-point equation to estimate the derivative for the following data at $x = 2.0$:

y	4.24	5.00	5.84
x	1.8	2.0	2.2

6 Use the three-point equation to estimate the derivative for the following data at $x = 1.9$:

y	3.41	3.57	3.76
x	1.8	1.9	2.0

7 Use the three-point equation with $h = 0.1$ to obtain a value for the derivative of the function $y = e^x$ at $x = 0$. Compare the result with that obtained analytically.

4.3.1 Taylor's theorem

Taylor's theorem (equation [7]) for $x = a + h$ is

$$f(a+h) = f(a) + hf'(a) + \frac{h^2}{2}f''(a) + \frac{h^3}{3!}f'''(a) + \dots$$

With $x = a - h$ the theorem gives

$$f(a+h) = f(a) - hf'(a) + \frac{h^2}{2}f''(a) - \frac{h^3}{3!}f'''(a) + \dots$$

Thus

$$f(a+h) - f(a-h) = 2hf'(a) + \frac{2h^3}{6}f'''(a) + \dots$$

Hence

$$f'(a) = \frac{f(a+h) - f(a-h)}{2h} - \frac{1}{6}h^2 f'''(a) - \dots \qquad [12]$$

The first term is the gradient given by the three-point equation. Thus the error is proportional to h^2. With the two-point equation the error is proportional to h. If we have $h \ll 1$ then we must have $h^2 \ll h$ and so the three-point equation is more accurate than the two-point equation.

4.4 Five-point equation

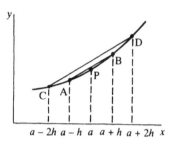

Fig. 4.3 Five-point equation

It seems reasonable to expect that the more points we use in the derivation of the derivative then the more accurate will be the result. Consider the situation shown in figure 4.3 where we have five points. We have two chords drawn symmetrically about the point P, one chord with an interval of $2h$ and the other with $4h$.

To see the relationship between these two chords, consider simple polynomial functions. If we had the function $y = x^3$ then drawing a chord AB on the graph of that function would give a chord, by equation [11], with a gradient of

$$\text{gradient}_{2h} = \frac{(a+h)^3 - (a-h)^3}{2h} = \frac{6a^2h + 2h^3}{2h}$$

$$= 3a^2 + h^2$$

But $dy/dx = 3x^2$ and so has the value $3a^2$ at $x = a$. Thus the gradient of the chord is

$$\text{gradient}_{2h} = \left(\frac{dy}{dx}\right)_{x=a} + h^2$$

If we had the function $y = x^4$ then drawing the chord on the graph of the function would give a chord with a gradient of

$$\text{gradient}_{2h} = \frac{(a+h)^4 - (a-h)^4}{2h} = \frac{8a^3h + 8ah^3}{2h}$$

$$= 4a^3 + 4ah^2$$

But $dy/dx = 4x^3$ and so has the value $4a^3$ at $x = a$. The gradient of the chord is thus

$$\text{gradient}_{2h} = \left(\frac{dy}{dx}\right)_{x=a} + 4ah^2$$

In general, with a polynomial, we have a chord gradient of the form

$$\text{gradient}_{2h} = \left(\frac{dy}{dx}\right)_{x=a} + Ah^2 \qquad [13]$$

where A is a constant.

Now, consider the chord CD. For the function $y = x^3$ we now have

$$\text{gradient}_{4h} = \frac{(a+2h)^3 - (a-2h)^3}{4h} = 3a^2 + 4h^2$$

In general we have

$$\text{gradient}_{4h} = \left(\frac{dy}{dx}\right)_{x=a} + 4Ah^2 \qquad [14]$$

where A is the constant in equation [13]. Hence, eliminating Ah^2 from the two equations [13] and [14],

$$\text{gradient}_{2h} = \left(\frac{dy}{dx}\right)_{x=a} + \frac{1}{4}\left[\text{gradient}_{4h} - \left(\frac{dy}{dx}\right)_{x=a}\right]$$

and thus

$$\left(\frac{dy}{dx}\right)_{x=a} = \frac{1}{3}\left[4(\text{gradient}_{2h}) - (\text{gradient}_{4h})\right] \qquad [15]$$

This equation relies on five points and is referred to as the *five-point equation*.

There is an alternative way of writing equation [15]. If we substitute the equations for the gradients in [11] then we have

$$\left(\frac{dy}{dx}\right)_{x=a} = \frac{4}{3}\frac{f(a+h) - f(a-h)}{2h} - \frac{1}{3}\frac{f(a+2h) - f(a-2h)}{4h}$$

$$= \frac{8f(a+h) - 8f(a-h) - f(a+2h) + f(a-2h)}{12h}$$

$$= \frac{-f(a+2h) + 8f(a+h) - 8f(a-h) + f(a-2h)}{12h} \qquad [16]$$

Example

Use the five-point equation to determine the derivative at $x = 0.5$ of the function described by the following data:

y	0.955	0.921	0.878	0.825	0.765
x	0.3	0.4	0.5	0.6	0.7

Taking $2h = 0.2$ then the chord between the points $x = 0.5 - 0.2$ and $x = 0.5 + 0.2$ gives a chord with a gradient of

$$\text{gradient}_{4h} = \frac{0.765 - 0.955}{2 \times 0.2} = -0.475$$

Taking points at $x = 0.5 - 0.1$ and $x = 0.5 + 0.1$ gives a chord with a gradient of

$$\text{gradient}_{2h} = \frac{0.825 - 0.921}{2 \times 0.1} = -0.480$$

Using equation [15],

$$\frac{dy}{dx} = \tfrac{1}{3}[4(-0.480) - (-0.475)] = -0.482$$

Review problems

8 Use the five-point equation to determine the derivative at $x = 0.5$ of the function described by the following data:

| y | 0.296 | 0.389 | 0.479 | 0.565 | 0.644 |
| x | 0.3 | 0.4 | 0.5 | 0.6 | 0.7 |

9 Use the five-point equation to determine the derivative at $x = 0.5$ of the function described by the following data:

| y | 0.39 | 0.56 | 0.75 | 0.96 | 1.19 |
| x | 0.3 | 0.4 | 0.5 | 0.6 | 0.7 |

4.4.1 Taylor's theorem

Taylor's theorem (equation [7]) for $x = a + h$ is

$$f(a+h) = f(a) + hf'(a) + \frac{h^2}{2}f''(a) + \frac{h^3}{3!}f'''(a) + \frac{h^4}{4!}f^{iv}(a)$$

$$+ \frac{h^5}{5!}f^{v}(a) + \dots$$

With $x = a - h$ the theorem gives

$$f(a-h) = f(a) - hf'(a) + \frac{h^2}{2}f''(a) - \frac{h^3}{3!}f'''(a) + \frac{h^4}{4!}f^{iv}(a)$$

$$- \frac{h^5}{5!}f^{v}(a) + \dots$$

Hence

$$f(a+h) - f(a-h) = 2hf'(a) + \frac{2h^3}{3!}f'''(a) + \frac{2h^5}{5!}f^v(a) + \ldots$$

and so

$$f'(a) = \frac{f(a+h) - f(a-h)}{2h} - \frac{1}{6}h^2 f'''(a) - \frac{1}{120}h^4 f^v(a) + \ldots$$

[17]

Taylor's theorem (equation [7]) for $x = a + 2h$ is

$$f(a+2h) = f(a) + 2hf'(a) + 2h^2 f''(a) + \frac{8h^3}{3!}f'''(a)$$

$$+ \frac{16h^4}{4!}f^{iv}(a) + \frac{32h^5}{5!}f^v(a) + \ldots$$

With $x = a - 2h$ the theorem gives

$$f(a-2h) = f(a) - 2hf'(a) + 2h^2 f''(a) - \frac{8h^3}{3!}f'''(a)$$

$$+ \frac{16h^4}{4!}f^{iv}(a) - \frac{32h^5}{5!}f^v(a) + \ldots$$

Thus

$$f(a+2h) - f(a-2h) = 4hf'(a) + \frac{16h^3}{3!}f'''(a)$$

$$+ \frac{64h^5}{5!}f^v(a) + \ldots$$

and so

$$f'(a) = \frac{f(a+2h) - f(a-2h)}{4h} - \frac{4}{6}h^2 f'''(a) - \frac{16}{120}h^4 f^v(a) + \ldots$$

[18]

We can eliminate the h^2 term, and so improve the accuracy, by subtracting equation [18] from four times equation [17] to give

$$4f'(a) - f'(a) = 4\left[\frac{f(a+h) - f(a-h)}{2h}\right]$$

$$-\left[\frac{f(a+2h) - f(a-2h)}{4h}\right] + \frac{1}{10}h^4 f^v(a) + \ldots$$

Thus

$$f'(a) = \frac{1}{3}\left[4(\text{gradient}_{2h}) - (\text{gradient}_{4h})\right] + \frac{1}{30}h^4 f^v(a) + \dots$$

$$[19]$$

This is the five-point equation given in equation [15], with the error proportional to h^4.

4.5 Commonly used approximations

Approximations that are commonly used to obtain derivatives numerically are:

Three-point equation:

$$f'(a) \approx \frac{f(a+h) - f(a-h)}{2h}$$

with error proportional to h^2

Four-point equation:

$$f'(a) = \frac{-f(x+2h) + 6f(x+h) - 3f(x) - 2f(x-h)}{6h}$$

with error proportional to h^3

Five-point equation:

$$f'(x) = \frac{-f(a+2h) + 8f(a+h) - 8f(a-h) + f(a-2h)}{12h}$$

with error proportional to h^4

Review problems

10 Determine, using the four-point equation, the value at $x = 0.2$ of $d(\sin x)/dx$. Use $h = 0.1$.
11 The following table gives the distance–time data for an object. Determine, using the four-point equation, the velocity at a distance of 54 mm.

Time in s	0.012	0.024	0.036	0.048
Distance in mm	19	54	100	168

4.5.1 Higher derivatives

In the same way as Taylor's theorem was used to obtain the first derivative then it can be used for second, and higher-order derivatives. For example, the three-point equation for the second-order derivative at $x = a$ is

$$f''(x) = \frac{f(a+h) - 2f(a) + f(a-h)}{h^2}$$ [20]

with error proportional to h^2

This equation can be derived as follows. Taylor's theorem gives for the value of a function $y = f(x)$ at $x = a + h$,

$$f(a+h) = f(a) + hf'(a) + \frac{h^2}{2}f''(a) + \dots$$ [21]

and at $x = a - h$,

$$f(a-h) = f(a) - hf'(a) + \frac{h^2}{2}f''(a) + \dots$$ [22]

Adding equations [21] and [22] gives

$$f(a+h) + f(a-h) = 2f(a) + h^2 f''(a) + \dots$$

Hence, with rearrangement, equation [20] is obtained.

Review problems

12 Determine, using three-point equations, the values at $x = 2$ of $d(\ln x)/dx$ and $d^2(\ln x)/dx^2$. Let $h = 0.1$.

Further problems

13 Determine the polynomial series which gives an approximate value for the function $y = f(x)$ if at $x = 0$ we have the values $y = 0$, $dy/dx = 3$, $d^2y/dx^2 = 4$ and $d^3y/dx^3 = -6$.

14 Determine the polynomial series which gives an approximate value for the function $y = f(x)$ if at $x = 2$ we have the values $y = 2$, $dy/dx = 1$, $d^2y/dx^2 = 6$ and $d^3y/dx^3 = 12$.

15 Using the two-point equation, determine the derivative of the function $y = \ln x$ at $x = 3$. Take an interval of $h = 0.01$.

16 Using the three-point equation, determine the derivative of the function $y = x\,e^x$ at $x = 1$. Take an interval of $h = 0.01$.

17 Determine the value of dy/dx for the function $y = x^2 + x + 1$ at $x = 0$, using (a) the two-point equation, (b) the three-point equation and (c) the four-point equation. Use $h = 0.1$.

18 Determine the value of dy/dx for the function $y = 2 \sin x$ at $x = 1$, using (a) the two-point equation, (b) the three-point equation. Use values of the interval $h = 0.1$, 0.2 and 0.3.

19 Determine the value of d^2y/dx^2 for the function $y = 2 \sin x$ at $x = 1$, using the three-point equation. Use $h = 0.1$.

20 Using the three-point equation, determine the value of d^2y/dx^2 at $y = 1.8$ for the following data:

y	1.712	1.901	2.236
x	1.7	1.8	1.9

21 Using Taylor's theorem, show that the error in the four-point equation is proportional to h^4.

5 Introducing integration

5.1 The concept of integration

There are two distinct ways we can approach the concept of *integration*. One is as the inverse of differentiation and the other is as an area under a graph of a function. The first leads to methods of evaluating integrals while the second leads to many applications in engineering. The term *integral* is used for the outcome of integration. This chapter introduces the principles of integration, with chapter 6 developing the rules and techniques of integration and chapters 7, 8 and 9 giving applications.

5.1.1 Integration as the inverse of differentiation

If we have a function $y = x^2$ then we can differentiate it to obtain $dy/dx = 2x$. We can define *integration* as the mathematical process which reverses differentiation, i.e. finds the function in this case which on being differentiated gives $2x$. Thus integrating $2x$ should give us x^2. However, the derivative of $x^2 + 3$ is also $2x$. Thus in the integration of $2x$ we are not sure whether there is a constant term or not. Hence a constant C has to be added to the result. Thus the outcome of the integration of x^2 is $2x + C$. The integral in such circumstances is, because of the constant, referred to as the *indefinite integral*.

Really what we are doing in the above process of integration is finding y in terms of x when given the values of dy/dx, i.e. the gradient of y against x, in terms of x. Figure 5.1 illustrates this. Thus in figure 5.1(a) we have a graph of the function $2x$. Integration of this results in a graph of the integral which is x^2 and which has gradients of $2x$.

We can indicate the process of integration by the use of the symbols $\int$ and dx. The $\int$ sign indicates that integration is to be carried out and the dx that x is the variable we are integrating with respect to. Thus the integration referred to above can be written as

$$y = \int 2x \, dx = x^2 + C$$

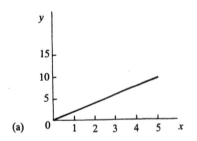

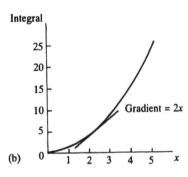

Fig. 5.1 Integrating $y = 2x$

In general

$$\int \frac{d}{dx}\{f(x)\}\,dx = f(x) + C \qquad\qquad [1]$$

Example

By considering integration as the reverse of differentiation, what is the integral of $3x^2$?

We have to find what function when differentiated gives $3x^2$. If we differentiate x^3 we obtain $3x^2$. If we differentiate $x^3 + C$ we obtain $3x^2$. Thus the integral of $3x^2$ is $x^3 + C$.

Example

Current i is the rate of movement of charge q with time t. The current through a capacitor in a circuit varies with time according to

$$i = 2\,e^{-3t}$$

Write an integral equation for the charge.

Since we have

$$\frac{dq}{dt} = 2\,e^{-3t}$$

then the reverse of this is

$$q = \int 2\,e^{-3t}\,dt$$

Review problems

1 By considering integration as the reverse of differentiation, determine the integrals of: (a) $3x^2 + 2x$, (b) $\sin x$, (c) $\sec^2 x$.
2 The current i through a capacitor depends on the rate of change of the voltage v across it, being given by

$$i = C\frac{dv}{dt}$$

where C is the capacitance. Write an integral equation for the voltage.
3 The velocity v of an object, i.e. the rate at which the distance x is being covered with respect to time t, is given by

$$v = \frac{dx}{dt} = u + at$$

where u and a are constants. Write an integral equation for the distance.

5.1.2 Integration as a series summation

Consider a moving object and its graph of velocity v against time t (figure 5.2). At the time $t = 0$ the object has zero velocity. The

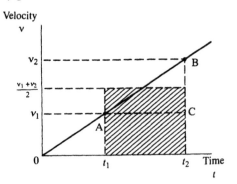

Fig. 5.2 Velocity–time graph

velocity increases as the time increases. The distance travelled x during some time interval is the product of the average speed during that interval multiplied by the time interval, i.e.

distance travelled = (average speed) × (time interval)

Thus for the time interval t_1 to t_2, when the velocity changes from v_1 to v_2, the distance travelled x is

$$x = \frac{v_2 + v_1}{2} \times (t_2 - t_1)$$

On the graph of velocity against time, the time interval is marked and the average velocity indicated. But the product of the average velocity and the time interval is the marked area. This is the same as the area under the graph line between A and B, i.e. the area of the rectangle between AC of $v_1(t_2 - t_1)$ and the time axis plus the area of the triangle ABC of $\frac{1}{2}(v_2 - v_1)(t_2 - t_1)$.

$$v_1(t_2 - t_1) + \tfrac{1}{2}(v_2 - v_1)(t_2 - t_1) = (v_1 + \tfrac{1}{2}v_2 - \tfrac{1}{2}v_1)(t_2 - t_1)$$

$$= \frac{v_1 + v_2}{2} \times (t_2 - t_1)$$

Thus the distance travelled is the area under the velocity–time graph.

We can represent the area under the velocity–time graph as the sum of a number of equal width strip areas. Figure 5.3 shows such strips. If t is the value of the time at the centre of a strip of

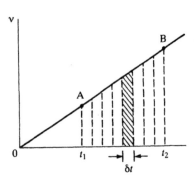

Fig. 5.3 Velocity–time graph

width δt and v the velocity at this time, then a strip has an area of $v\,\delta t$. Thus the area under the graph between A and B is equal to the sum of the areas of all such strips between the times t_1 and t_2, i.e.

distance travelled = area under graph between t_1 and t_2

= sum of the areas of all the strips between t_1 and t_2

We can write this summation as

$$x = \sum_{t=t_1}^{t=t_2} v\,\delta t$$

If we make δt very small, i.e. let $\delta t \rightarrow 0$, then we denote it by dt. The sum is then the sum of a series of very narrow strips and is written as

$$x = \lim_{\delta t \rightarrow 0} \sum_{t=t_1}^{t=t_2} v\,\delta t = \int_{t_1}^{t_2} v\,dt$$

The sign $\int$ is an "S" for summation and the t_1 and t_2 are said to be the limits of the range of the variable t. Here x is the *integral* of the function v between the limits t_1 and t_2. The process of obtaining x in this way is termed *integration*. Because the integration is between specific limits it is referred to as a *definite integral*. In general we write for the integration of a function $f(x)$ between limits, i.e. the determination of the area under the graph of $f(x)$ between limits,

$$\text{area} = \lim_{\delta x \rightarrow 0} \sum_{x=x_1}^{x=x_2} f(x)\,\delta x = \int_{x_1}^{x_2} f(x)\,dx \qquad [2]$$

We can show that the definitions of integration in terms of the inverse of differentiation and the area under a graph describe the same concept. Suppose we increase the area under a graph of the function $y = f(x)$ against x by one strip. Then

increase in area $\delta A = y\,\delta x$

So

$$\frac{\delta A}{\delta x} = y$$

In the limit as $\delta x \rightarrow 0$ then we can write dA/dx and so

$$\frac{dA}{dx} = y = f(x)$$

In section 5.1.1, integration was defined as the inverse of differentiation. Thus, by that definition, the integration of the above equation gives the area A.

$$A = \int f(x)\, dx$$

This is an indefinite integral, which is the same as that given by the definition for integration as the area under a graph when limits are imposed.

Example

By considering the area under a graph of $y = x + 1$, determine the integral

$$\int_{-3}^{4} (x + 1)\, dx$$

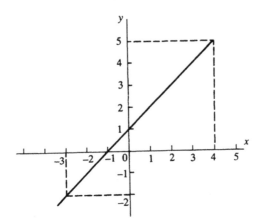

Fig. 5.4 Example

Figure 5.4 shows the graph. The integral is the area between the values of x of 4 and -3. The area under the graph between $x = 0$ and $x = 4$ is that of a rectangle 4×1 plus a triangle $\frac{1}{2}(4 \times 4)$ and so is $+12$ square units. The area between $x = -1$ and 0 is that of a triangle $\frac{1}{2}(1 \times 1) = 0.5$ square units and the area between $x = -1$ and $x = -3$ is a triangular area below the axis and so is negative and given by $-\frac{1}{2}(2 \times 2) = -2$ square units. Hence the total area under the graph is $+12 + 0.5 - 2 = 10.5$ square units.

Review problems

4   By considering the area under a graph of $y = x + 5$, determine the integral

$$\int_{-3}^{2} (x + 5)\, dx$$

5   By considering the area under a graph of $y = \sin x$, determine the integral

$$\int_{-\pi}^{\pi} \sin x \, dx$$

6   The work done w is the area under a graph of force F against displacement x in the direction of the force. Determine the work done when $F = 2x$ and x changes from 0 to 2, i.e.

$$\int_{0}^{2} 2x \, dx$$

5.1.3 Indefinite and definite integrals

An *indefinite integral* has no limits and the result has a constant of integration. Integration between specific limits gives a *definite integral*.

Consider the integration of the function $y = 2x$. This has no specified limits and so is an indefinite integral, with the solution (as the function which differentiated would give $2x$) of

$$\int 2x \, dx = x^2 + C$$

Now consider the area under the graph of $y = 2x$ between the limits of $x = 1$ and $x = 3$. We can write this as

$$\int_{1}^{3} 2x \, dx = [x^2 + C]_{1}^{3}$$

The square brackets round the $x^2 + C$ is used to indicate that we have to impose the limits of 3 and 1 on it. Thus

$$\int_{1}^{3} 2x \, dx = (9 + C) - (1 + C) = 8$$

The constant term C vanishes when we have a definite integral.

It is important when writing the limits of the integral that they are the right way round. In the above case we are determining the area between $x = 1$ and $x = 3$. Suppose we wanted the area between $x = 3$ and $x = 1$. The integral would now be

$$\int_{3}^{1} 2x \, dx = [x^2 + C]_{3}^{1} = (1 + C) - (9 + C) = -8$$

Interchanging the limits changes the sign of the integral.

In the above discussion of definite integrals we have assumed that the function is continuous between the limits. Consider the function shown in figure 5.5. If we want to integrate that function between the limits $x = 0$ and $x = 4$ then it is simplest if we consider

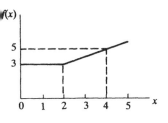

Fig. 5.5 A function which can be considered in two segments

the area under the graph in two pieces, evaluating the part of the area between the limits $x = 0$ and $x = 2$ and then adding it to the part between the limits $x = 2$ and $x = 4$. Thus we have

$$\int_0^4 f(x)\,dx = \int_0^2 3\,dx + \int_2^4 (x+1)\,dx$$

There are, however, two cases when the evaluation of a definite integral requires special care. These are when:

1   one or both of the limits of the integral are infinite,

2   the function being integrated becomes infinite at one or more points between the limits.

Our definition of the definite integral in terms of the area under a graph of a function only really works if the function does not go off to infinity in the interval between the specified limits or at the limits. If either of the above conditions occurs then the integral is said to be an *improper integral*. Such integrals can, however, often still be evaluated.

Consider the definite integral

$$\int_1^\infty \frac{1}{x^2}\,dx$$

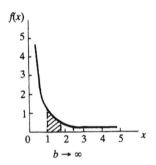

Fig. 5.6 $f(x) = 1/x^2$

This has one limit which is set at infinity so it is an improper integral. Figure 5.6 shows a graph of the function. We can attempt to evaluate such an integral by considering the integral with a finite limit b instead of infinity, evaluating it with this limit and then considering what happens to the solution in the limit as b tends to the infinite value. Thus

$$\int_1^\infty \frac{1}{x^2}\,dx = \lim_{b \to \infty}\left[\int_1^b \frac{1}{x^2}\,dx\right] = \lim_{b \to \infty}\left[-\frac{1}{x}+C\right]_1^b$$

$$= \lim_{b \to \infty}\left(-\frac{1}{b}+1\right) = 1$$

Hence we define the value of the improper integral

$$\int_a^\infty f(x)\,dx$$

as being given by

$$\int_a^\infty f(x)\,dx = \lim_{b \to \infty}\int_a^b f(x)\,d$$

Not all improper integrals will converge in this way to a finite value. They only converge if the area under the graph within the limits can be considered to have a finite value. For some functions the area is infinite.

With improper integrals that go off to infinity at some point between the limits then we can consider the area under the graph of the function in two pieces, from the lower limit up to the ordinate at which the function goes off to infinity and then from that ordinate to the upper limit. Each part can then be treated in the way outlined above.

Example

Determine the value of the improper integral

$$\int_1^\infty e^{-x}\, dx$$

Replacing the infinite limit by one of b we have (see table 5.1, page 102, for the integral of the exponential term)

$$\int_1^b e^{-x}\, dx = [-e^{-x}]_1^b = -e^{-b} - (-e^{-1}) = \frac{1}{e} - e^{-b}$$

As b tends to infinity then the integral tends to the value of $1/e$.

Review problems

7   Determine the values, if they exist, of the following integrals:

(a) $\int_0^2 2x\, dx$, (b) $\int_1^2 2x\, dx$, (c) $\int_2^1 2x\, dx$, (d) $\int_{-1}^2 2x\, dx$,

(e) $\int_0^\infty 2x\, dx$

.2 Integrals of common unctions

The integrals of function can be determined by considering what function will give the function when differentiated.

5.2.1 Integration of x^n

Consider the integration of

$$\int x^n\, dx$$

Considering integration as the inverse of differentiation, the question becomes as to what function gives x^n when differentiated. The derivative of x^{n+1} is $(n + 1)x^n$ (see item 2, table 1.1). Thus the derivative of $x^{n+1}/(n + 1)$ is x^n. Hence

$$\int x^n \, dx = \frac{x^{n+1}}{n+1} + C \qquad\qquad [3]$$

This is true for positive, negative and fractional values of n other than $n = -1$.

Consider the integral of x^{-1}, i.e.

$$\int \frac{1}{x} \, dx$$

The derivative of $\ln x$ is $1/x$ (see table 1.1, item 4). Thus

$$\int \frac{1}{x} \, dx = \ln x + C \qquad\qquad [4]$$

This applies if x is positive, i.e $x > 0$. If x is negative, i.e. $x < 0$, then the integral of $1/x$ in such a situation is *not* $\ln x$. This is because we cannot have the logarithm of a negative number as a real quantity (see the Appendix); $\ln x$ is only defined for $x > 0$. For this reason you will often find equation [4] written as

$$\int \frac{1}{x} \, dx = \ln |x| + C$$

to indicate that we are only ever dealing with positive values for the logarithm.

Example

Evaluate the integrals:

(a) $\int x^8 \, dx$, (b) $\int x^{1/2} \, dx$, (c) $\int x^{-3} \, dx$, (d) $\int x^{-1} \, dx$,

(e) $\int_0^2 x \, dx$, (f) $\int_0^2 x^{-3} \, dx$, (g) $\int_1^\infty x^{-2} \, dx$

(a) Using equation [3],

$$\int x^8 \, dx = \frac{x^9}{9} + C$$

(b) Using equation [3],

$$\int x^{1/2} \, dx = \frac{x^{3/2}}{3/2} + C$$

(c) Using equation [3],

$$\int x^{-3} \, dx = \frac{x^{-2}}{-2} + C$$

(d) Using equation [4],

$$\int x^{-1}\, dx = \ln x + C$$

(e) Using equation [3],

$$\int_0^2 x\, dx = \left[\tfrac{1}{2}x^2 + C \right]_0^2 = 2$$

(f) Using equation [3],

$$\int_0^2 x^{-3}\, dx = \left[-\tfrac{1}{2}x^{-2} + C \right]_0^2$$

But this implies an infinite answer. The integral is an improper integral, it going off to infinity at $x = 0$. We can try the method outlined in the section 5.1.3 to see if an answer is possible. Thus

$$\int_0^2 x^{-3}\, dx = \lim_{b \to \infty} \left[\int_b^2 x^{-3}\, dx \right] = \lim_{b \to \infty} \left[-\tfrac{1}{2}x^{-2} + C \right]_b^2$$

$$= \lim_{b \to \infty} \left(-\frac{1}{8} + \frac{1}{2b^2} \right)$$

The improper integral gives a result which diverges and so we cannot give an answer.

(g) This is an improper integral and thus we use the method outlined in section 5.1.3 to see if an answer is possible.

$$\int_1^\infty x^{-2}\, dx = \lim_{b \to \infty} \left[\int_1^b x^{-2}\, dx \right] = \lim_{b \to \infty} \left[-x^{-1} + C \right]_1^b$$

$$= \lim_{b \to \infty} \left(-\frac{1}{b} + 1 \right) = 1$$

Review problems

8 Determine the integrals:

(a) $\int x^3\, dx$, (b) $\int x^{-7}\, dx$, (c) $\int x^{0.3}\, dx$, (d) $\int x^{1/3}\, dx$,

(e) $\int_1^2 x\, dx$, (f) $\int_0^1 x^3\, dx$, (g) $\int_1^\infty \frac{1}{x}\, dx$

5.2.2 Common functions

Table 5.1 shows some commonly encountered functions and their integrals.

Table 5.1 Integrals

	Function $f(x)$	$\int f(x)\,dx$
1	k, a constant	$kx + C$
2	x^n	$\dfrac{x^{n+1}}{n+1} + C$, for $n \neq -1$
3	e^{at}	$\dfrac{1}{a}e^{at} + C$
4	$\dfrac{1}{x}$	$\ln x + C$ for $x > 0$
5	$\cos ax$	$\dfrac{1}{a}\sin ax + C$
6	$\sin ax$	$-\dfrac{1}{a}\cos ax + C$
7	$\tan ax$	$\dfrac{1}{a}\ln(\sec ax) + C$
8	$\sec^2 ax$	$\dfrac{1}{a}\tan ax + C$
9	$\sec ax$	$\dfrac{1}{a}\{\ln(\operatorname{cosec} ax - \cot ax)\} + C$
10	$\operatorname{cosec} ax$	$\dfrac{1}{a}\{\ln(\sec ax + \tan ax)\} + C$
11	$\cot ax$	$\dfrac{1}{a}\{\ln(\sin ax)\} + C$
12	$\dfrac{1}{\sqrt{1-(ax)^2}}$ for $(ax)^2 < 1$	$\dfrac{1}{a}\sin^{-1} ax + C$
13	$\dfrac{1}{x\sqrt{(ax)^2-1}}$ for $(ax)^2 > 1$	$\dfrac{1}{a}\sec^{-1} ax + C$
14	$\dfrac{1}{1+(ax)^2}$	$\dfrac{1}{a}\tan^{-1} ax + C$
15	$\cosh ax$	$\dfrac{1}{a}\sinh ax + C$
16	$\sinh ax$	$\dfrac{1}{a}\cosh ax + C$
17	$\operatorname{sech}^2 ax$	$\dfrac{1}{a}\tanh ax + C$
18	$\dfrac{1}{\sqrt{(ax)^2+1}}$	$\dfrac{1}{a}\sinh^{-1} ax + C$
19	$\dfrac{1}{\sqrt{(ax)^2-1}}$ for $(ax)^2 > 1$	$\dfrac{1}{a}\cosh^{-1} ax + C$
20	$\dfrac{1}{1-(ax)^2}$ for $(ax)^2 < 1$	$\dfrac{1}{a}\tanh^{-1} ax + C$

Note: the integral of $1/x$ should be stated as $\ln|x| + C$, since a logarithm can only be of a positive quantity. For the same reason this also applies to items 7, 9, 10 and 11.

Example

Use table 5.1 to obtain the following integrals:

(a) $\int \cos 4x \, dx$, (b) $\int 6 \, dx$, (c) $\int e^{2x} \, dx$, (d) $\int \tan(2x-4) \, dx$

(a) Using item 5,

$$\int \cos 4x \, dx = \tfrac{1}{4} \sin 4x + C$$

(b) Using item 1,

$$\int 6 \, dx = 6x + C$$

(c) Using item 3,

$$\int e^{2x} \, dx = \tfrac{1}{2} e^{2x} + C$$

(d) Using item 7,

$$\int \tan(2x-4) \, dx = \tfrac{1}{2} \ln\{\sec(2x-4)\} + C$$

Example

Using table 5.1, determine the following integrals:

(a) $\int_{-2}^{4} e^{2x} \, dx$, (b) $\int_{0}^{\pi/3} \cos 2x \, dx$, (c) $\int_{-\infty}^{1} e^{2x} \, dx$

(a) Using item 3,

$$\int_{-2}^{4} e^{2x} \, dx = \left[\tfrac{1}{2} e^{2x} + C \right]_{-2}^{4} = \tfrac{1}{2} e^{8} - \tfrac{1}{2} e^{-4} = 1490$$

(b) Using item 5,

$$\int_{0}^{\pi/3} \cos 2x \, dx = \left[\tfrac{1}{2} \sin 2x + C \right]_{0}^{\pi/3} = \tfrac{1}{2} \sin 2\pi/3 - \tfrac{1}{2} \sin 0$$

$$= 0.433$$

(c) This is an improper integral. We can thus try the method given in section 5.1.3 to see if an answer is possible. Thus

$$\int_{-\infty}^{1} e^{2x} \, dx = \lim_{b \to -\infty} \left[\int_{b}^{1} e^{2x} \, dx \right] = \lim_{b \to -\infty} \left[\tfrac{1}{2} e^{2x} + C \right]$$

$$= \lim_{b \to -\infty} \left(\tfrac{1}{2} e^{2} - \tfrac{1}{2} e^{2b} \right) = \tfrac{1}{2} e^{2} - 0 = 3.69$$

Review problems

9 Use table 5.1 to obtain the following integrals:

(a) $\int 2\,dx$, (b) $\int e^{-2x}\,dx$, (c) $\int \sin 3x\,dx$, (d) $\int \cot 4x\,dx$,

(e) $\int \frac{1}{x}\,dx$, (f) $\int \cosh 2x\,dx$, (g) $\int \sec 5x\,dx$

10 Use table 5.1 to obtain the following integrals:

(a) $\int_1^3 5\,dx$, (b) $\int_0^{\pi/4} \cos 2x\,dx$, (c) $\int_1^2 e^x\,dx$, (d) $\int_0^2 \frac{1}{x}\,dx$,

(e) $\int_0^\infty e^{-x}\,dx$

11 The velocity v of an object is given by the equation

$$v = \frac{dx}{dt} = t$$

Hence determine how the distance x varies with time.

5.3 Integral equations

Consider an object moving in a straight line with a velocity v which is a function of time and increases with time t in the way described by the equation $v = at$, where a is a constant. Since we can write $v = dx/dt$ then we have

$$\frac{dx}{dt} = at$$

Now suppose we integrate both sides of the equation with respect to time t, we then have

$$\int \frac{dx}{dt}\,dt = \int at\,dt$$

and so

$$\int dx = \int at\,dt$$

Since the integral of $1\,dx$ is $x + C$, with C being a constant, then

$$x + C = \int at\,dt$$

When we integrate the right-hand side of the equation we will end up with another constant of integration. We can combine these two constants and so just write

$$x = \int at \, dt$$

This is just another way of looking at the equation $dx/dt = at$ and stating that the integral is the inverse of the derivative.

Review problems

12 The rate of decay dN/dt of N unstable nuclei in a radioactive sample is related to N by

$$\frac{dN}{dt} = -kN$$

where k is a constant. Write this as an integral equation.

13 The potential difference v across an inductor is related to the rate of change of current i through it by the equation

$$v = L\frac{di}{dt}$$

where L is a constant, the inductance. Write this as an integral equation.

Further problems

14 If $dy/dx = \cos x$, what is y? Consider this integration as the inverse of differentiation.

15 The velocity v of a moving object is related to the time t by the equation $v = 3 + 2t$. Write an integral equation for the displacement x of the object.

16 By considering the area under a graph of $y = 2x + 1$, determine the integral

$$\int_{1}^{3}(2x+1)\,dx$$

17 By considering the area under a graph of $y = \sin x$, determine the integral

$$\int_{0}^{2\pi} \sin x \, dx$$

18 Determine the following integrals:

(a) $\int x^9 \, dx$, (b) $\int x^{-5} \, dx$, (c) $\int \sqrt{x} \, dx$, (d) $\int \frac{1}{\sqrt{x}}\, dx$, (e) $\int 7 \, dx$,

(f) $\int e^{-3x} dx$, (g) $\int e^{x/2} dx$, (h) $\int \cos x \, dx$, (i) $\int \sec 4x \, dx$,

(j) $\int \sin 2x \, dx$, (k) $\int \cos 2x \, dx$, (l) $\int \cosh 2x \, dx$

19 Evaluate the following integrals:

(a) $\int_1^2 x^2 \, dx$, (b) $\int_0^{\pi/2} \sin x \, dx$, (c) $\int_{-1}^4 \frac{1}{x^2} dx$,

(d) $\int_{\pi/4}^{\pi/2} \cos 2x \, dx$, (e) $\int_0^2 \frac{1}{x} dx$

20 A moving iron ammeter has a current sensitivity given by

$$\frac{d\theta}{dI} = cI$$

where θ is the angular deflection of the pointer for a current I and c is a constant. Represent this as an integral equation.

21 The current i through a capacitor varies with time t according to the equation

$i = 3 \sin 50t$

Write an integral equation describing how the charge on the capacitor will vary with time.

22 The velocity v of an object varies with time t according to the equation

$v = 10t + 3$

Derive an integral equation relating the displacement x and time.

6 Techniques of integration

6.1 Basic rules

This chapter is about basic rules and techniques that are involved in integration. This opening section is about the basic rules with the following sections demonstrating techniques commonly used to rearrange functions so that they can be integrated. These are: substitutions, integration by parts and partial fractions. These rules and techniques are used in applications involving chapters 7, 8 and 9. This chapter can thus be regarded as introducing the rules and techniques illustrated with examples and problems for practice, these being:

1. Function multiplied by a constant (6.1.1.).
2. Sum of functions (6.1.2).
3. The technique of substitution (6.2).
4. The technique of integration by parts (6.3).
5. The technique of integration by partial fractions (6.4).

6.1.1 Multiplication by a constant

The derivative, for example, of $5x^2$ is 5 multiplied by the derivative of x^2, i.e. $10x$. The derivative of $af(x)$, where a is a constant, is (see section 2.1.2)

$$\frac{d}{dx}\{af(x)\} = a\frac{d}{dx}\{f(x)\}$$

Thus the integral of $af(x)$ is the constant a multiplied by the integral of $f(x)$, i.e.

$$\int af(x)\,dx = a\int f(x)\,dx \qquad [1]$$

Example

Evaluate the following integrals:

(a) $\int 4x^2 \, dx$, (b) $\int 2\cos x \, dx$, (c) $\int \frac{4}{x} \, dx$

(a) Using the above rule for the multiplication by a constant,

$$\int 4x^2 \, dx = 4 \int x^2 \, dx = \frac{4}{3}x^3 + C$$

(b) Using the rule for the multiplication by a constant,

$$\int 2\cos x \, dx = 2 \int \cos x \, dx = 2\sin x + C$$

(c) Using the rule for the multiplication by a constant.

$$\int \frac{4}{x} \, dx = 4 \int \frac{1}{x} \, dx = 4\ln x + C$$

Sometimes it is convenient to write this as $4 \ln x + \ln A$, where the constant C is written in terms of another constant A, i.e. $C = \ln A$. Then, combining the logarithms, the solution can be written as $\ln Ax^4$.

Review problems

1   Evaluate the following integrals:

(a) $\int 5x \, dx$, (b) $\int 2\sin 3x \, dx$, (c) $\int 3 e^{4x} \, dx$

2   When a capacitor is being charged, then the total work done in charging it from zero charge to a charge Q is given by

$$\text{work} = \int_0^Q \frac{q}{C} \, dq$$

with q being the charge on the capacitor at some instant and C a constant, the capacitance. Evaluate the integral.

6.1.2 Sum

The derivative of, for example, $x^2 + x$ is $2x + 1$, i.e. the sum of the derivatives of x^2 and x when considered separately. Since the derivative of the sum of a number of functions is the sum of the derivatives of each function when considered alone (see section 2.1.1), then the integral of the sum of a number of functions is the sum of their separate integrals.

$$\int [f(x) + g(x)]\, dx = \int f(x)\, dx + \int g(x)\, dx \qquad [2]$$

Example

Evaluate the integral

$$\int (4x^3 + 2x^2 - 5x + 1)\, dx$$

Using the summation rule in equation [2],

$$\int (4x^3 + 2x^2 - 5x + 1)\, dx = \int 4x^3\, dx + \int 2x^2\, dx - \int 5x\, dx + \int 1\, dx$$

Hence, integrating each element gives

$$\int (4x^3 + 2x^2 - 5x + 1)\, dx = x^4 + \tfrac{2}{3}x^3 - \tfrac{5}{2}x^2 + x + C$$

All the constants resulting from each integration have been incorporated into the one constant C.

Example

Evaluate the integral:

$$\int (x+4)^2\, dx$$

Functions often have to be rearranged into the standard integral forms before integration. Thus we can write the integral as

$$\int (x+4)^2\, dx = \int (x^2 + 8x + 16)\, dx$$

$$= \int x^2\, dx + 8 \int x\, dx + \int 16\, dx$$

$$= \frac{x^3}{3} + 8\frac{x^2}{2} + 16x + C$$

Review problems

3 Evaluate the following integrals:

(a) $\int (x^2 + 3x + 1)\, dx$, (b) $\int \left(x + \frac{1}{x}\right) dx$, (c) $\int (\sin 2x + \cos 2x)\, dx$,

(d) $\int (e^x + e^{-x})\, dx$, (e) $\int (x+2)^2\, dx$, (f) $\int \frac{x^3 - 2x}{4x}\, dx$,

(g) $\int \frac{(1+x)^2}{\sqrt{x}}\, dx$

Table 6.1 Substitutions

If $f(x)$ contains	Try the substitution
1 $ax + b$	$u = ax + b$
2 $\sqrt{ax + b}$	$u^2 = ax + b$
3 $\sqrt{a^2 - x^2}$	$x = a \sin \theta$ or $x = a \tanh u$
4 $\sqrt{a^2 + x^2}$	$x = a \tan \theta$ or $x = a \sinh u$
5 $a^2 + x^2$	$x = a \tan \theta$ or $x = a \sinh u$
6 $\sqrt{x^2 - a^2}$	$x = a \sec \theta$ or $x = a \cosh u$
7 Trigonometric functions	$s = \sin x$ or $c = \cos x$ or $t = \tan x$
8 Combination of trignonometric functions	$t = \tan \dfrac{x}{2}$

Note that, by using trigonometric relationships:

$$\sin x = \frac{2t}{1 + t^2} \quad \text{and} \quad \cos x = \frac{1 - t^2}{1 + t^2} \quad \text{and} \quad \tan x = \frac{2t}{1 - t^2}$$

$$dx = \frac{2}{1 + t^2} \, dt$$

9 Hyperbolic functions	$s = \sinh x$ or $c = \cosh x$ or $t = \tanh x$
10 Combination of hyperbolic functions	$t = \tanh \dfrac{x}{2}$

Note that, by using hyperbolic relationships:

$$\sinh x = \frac{2t}{1 - t^2} \quad \text{and} \quad \cosh x = \frac{1 + t^2}{1 - t^2} \quad \text{and} \quad \tanh x = \frac{2t}{1 + t^2}$$

$$dx = \frac{2}{1 - t^2} \, dt$$

Note: see the Appendix for details of trigonometric and hyperbolic relationships.

Hint: for (e), (f) and (g) rewrite the function in a form which enables you to use the standard integrals given in table 5.1.

4   The velocity v of an object as a function of time t is given by the equation

$$v = 2 + 3t$$

Determine the equation relating the displacement and time.

6.2 Integration by substitution   A very useful technique for evaluating of the integrals of many functions is *substitution*. This involves the substitution of another variable for part of the function, the variable chosen being one that enables the function to be put into a form for which standard integrals can be used. The variable chosen can take many forms. In this section we consider substitutions involving a simple algebraic function. Later sections will consider other forms of substitution.

Consider obtaining the integral of $(2x + 3)^2$. We can simplify this by making the substitution $u = 2x + 3$. Thus

$$\int (2x + 3)^2 \, dx = \int u^2 \, dx$$

We cannot, however, integrate a function of the variable u with respect to a different variable x. But $u = 2x + 3$ gives $du/dx = 2$ and so we can substitute $du/2$ for dx. Hence

$$\int (2x + 3)^2 \, dx = \int \frac{u^2}{2} \, du = \frac{u^3}{6} + C = \frac{1}{6}(2x + 3)^2 + C$$

The method involved in the substitution method is comparable to the chain rule in differentiation (see section 2.2).

The technique used can be summarised as:

1 Choose a substitution u equal to some function of the variable x. Usually it is best to choose the inner part of a composite function, such as the $(2x + 3)$ in the above example of the function $(2x + 3)^2$.
2 Evaluate the relationship between du and dx.
3 Rewrite the integral in terms of u and du.
4 Evaluate the resulting integral.
5 Back substitute to obtain the result in terms of x.

There are no rules determining which substitution should be used, the aim being to choose one which results in simplification of the integral. If the factor $(ax + b)$ appears in the function being integrated with respect to x then the substitution $u = ax + b$ is tried. If $\sqrt{(ax + b)}$ then $u^2 = ax + b$ might be chosen. See the following sections of this chapter for details of other substitutions and table 6.1 for a summary of possible substitutions.

There is a particularly useful rule that can be developed for some functions where the numerator is the derivative of the denominator, i.e. we have

$$\int \frac{f'(x)}{f(x)} \, dx$$

If we let $u = f(x)$ then $du/dx = f'(x)$ and so the integral can be written as

$$\int \frac{f'(x)}{f(x)} \, dx = \int \frac{f'(x)}{u} \times \frac{du}{f'(x)} = \int \frac{1}{u} \, du = \ln u + C = \ln\{f(x)\} + C$$

Thus we have

$$\int \frac{f'(x)}{f(x)} \, dx = \ln \{f(x)\} + C \qquad\qquad [3]$$

This can be illustrated by the following example,

$$\int \frac{2x}{x^2 + 1} \, dx = \ln(x^2 + 1) + C$$

This is because we have $f(x) = x^2 + 1$ and the derivative of this function is $2x$.

Example

Evaluate the integral

$$\int (3x + 1)^{1/2} \, dx$$

Let $u = 3x + 1$. Then $du/dx = 3$. Thus the integral can be written as

$$\int (3x + 1)^{1/2} \, dx = \int \frac{u^{1/2}}{3} \, du = \frac{u^{3/2}}{3 \times 3/2} + C$$

$$= \tfrac{2}{9}(3x + 1)^{3/2} + C$$

We could have tried a different substitution. Let $u^2 = 3x + 1$. Then $2u \, du/dx = 3$. Thus the integral can be written as

$$\int (3x + 1)^{1/2} \, dx = \int u \left(\frac{2u \, du}{3} \right) = \frac{2u^3}{3 \times 3} + C$$

$$= \tfrac{2}{9}(3x + 1)^{3/2} + C$$

Example

Evaluate the integral

$$\int x(2x^2 + 1)^2 \, dx$$

Let $u = 2x^2 + 1$. Then $du/dx = 4x$. Hence the integral becomes

$$\int x(2x^2 + 1)^2 \, dx = \int xu^2 \frac{du}{4x} = \frac{u^3}{4 \times 3} + C$$

$$= \tfrac{1}{12}(2x^2 + 1)^3 + C$$

Example

Evaluate the integral

$$\int 2x \sin x^2 \, dx$$

Let $u = x^2$. Then $du/dx = 2x$. Hence the integral becomes

$$\int 2x \sin x^2 \, dx = \int \sin u \, du = -\cos u + C$$

$$= -\cos x^2 + C$$

Review problems

5 Evaluate, by substitution, the following integrals:

(a) $\int \cos(2x + 5) \, dx$, (b) $\int (2x + 4)^7 \, dx$, (c) $\int e^{2x-1} \, dx$,

(d) $\int 2x(4x^2 + 1)^5 \, dx$, (e) $\int \dfrac{x}{3x^2 + 2} \, dx$, (f) , $\int (2 - 3x)^3 \, dx$,

(g) $\int x^2(x^3 + 2)^{1/2} \, dx$, (h) $\int \dfrac{x+1}{x^2 + 2x + 2} \, dx$, (i) $\int \dfrac{x^2}{\sqrt{x+2}} \, dx$,

(j) $\int \dfrac{3x^2}{x^3 + 2} \, dx$, (k) $\int \dfrac{1}{x+1} \, dx$, (l) $\int \dfrac{2x+3}{x^2 + 3x + 2} \, dx$

6 The potential difference V between two parallel wires is given by

$$V = -\int_{D-a}^{a} \frac{Q}{2\pi\varepsilon_r\varepsilon_0}\left(\frac{1}{x} + \frac{1}{D-x}\right) dx$$

where a is the radius of the wires and D the distances between the wire centres. Q, ε_r and ε_0 are constants. Evaluate the integral.

7 The rate at which a hot object cools $d\theta/dt$ is proportional to the difference in temperatures between that of the hot body θ and its surroundings θ_s, i.e.

$$\frac{d\theta}{dt} = k(\theta - \theta_s)$$

Hence determine an expression indicating how θ varies with time t.

6.2.1 Trigonometric integrals

In integrating trigonometric functions, the use of trigonometric relationships (see the Appendix) may put functions into a form which can be readily integrated. For example, the integral of $\sin^2 x$ may be obtained by replacing it by $\frac{1}{2}(1 - \cos 2x)$. This can then be directly integrated.

With some functions, substitutions made with trigonometric functions may be used to enable integration to be carried out. The substitutions are typically $s = \sin x$, $c = \cos x$ or $t = \tan x$. The following examples illustrate this.

Integrals of the form

$$\int \cos^m x \sin^n x \, dx \tag{4}$$

often occur in engineering, e.g. in the Fourier series representation of non-sinusoidal signals. When one of the powers is odd and the other even we can use a simple substitution to solve the integral. If m is odd we let $s = \sin x$ and use $\cos^2 x = 1 - \sin^2 x$. If n is odd we let $c = \cos x$ and use $\sin^2 x = 1 - \cos^2 x$.

Consider for example

$$\int \cos^5 x \sin^2 x \, dx$$

Let $s = \sin x$, then $ds/dx = \cos x$ and so the integral becomes

$$\int \cos^5 x \sin^2 x \, dx = \int \cos^5 x \times s^2 \times \frac{ds}{\cos x} = \int s^2 \cos^4 x \, ds$$

Since $\sin^2 x + \cos^2 x = 1$ (see the Appendix) then

$$\cos^4 x = (\cos^2 x)^2 = (1 - \sin^2 x)^2 = (1 - s^2)^2$$

Thus

$$\int \cos^5 x \sin^2 x \, dx = \int s^2 (1 - s^2)^2 \, ds = \int (s^2 - 2s^4 + s^6) \, ds$$

and hence

$$\int \cos^5 x \sin^2 x \, dx = \frac{s^3}{3} - \frac{2s^5}{5} + \frac{s^7}{7} + C$$

$$= \frac{1}{3} \sin^3 x - \frac{2}{5} \sin^5 x + \frac{1}{7} \sin^7 x + C$$

When both of the powers m and n in integral [4] are even or both odd and small then the above method cannot be used. The integral is then found by using such relationships (for a fuller list, see the Appendix) as:

$$\sin^2 x = \frac{1}{2}(1 - \cos 2x)$$

$$\cos^2 x = \frac{1}{2}(1 + \cos 2x)$$

$$\sin x \cos x = \tfrac{1}{2}\sin 2x$$

Thus, for example, if we have

$$\int \cos^4 x \sin^2 x \, dx$$

then we can simplify the integral by using the relationships

$$\cos^2 2x = \tfrac{1}{2}(1 + \cos 2x)$$

$$\sin^2 2x = \tfrac{1}{2}(1 - \cos 2x)$$

The integral then becomes

$$\int \cos^4 x \sin^2 x \, dx = \int \tfrac{1}{4}(1 + \cos 2x)^2 \tfrac{1}{2}(1 - \cos 2x) \, dx$$

$$= \int \tfrac{1}{8}(1 + \cos 2x - \cos^2 2x - \cos^3 2x) \, dx$$

We can put these terms into the standard form by using

$$\cos^2 2x = \tfrac{1}{2}(1 + \cos 4x)$$

$$\cos^3 2x = \cos^2 2x \cos 2x = (1 - \sin^2 2x)\cos 2x$$

Hence

$$\int \cos^4 x \sin^2 x \, dx = \int \tfrac{1}{8} \, dx + \int \tfrac{1}{8} \cos 2x - \int \tfrac{1}{16} \, dx - \int \tfrac{1}{16} \cos 4x \, dx$$

$$- \int \tfrac{1}{8} \cos 2x + \int \tfrac{1}{8} \sin^2 2x \cos 2x$$

The integral

$$\int \tfrac{1}{8} \sin^2 2x \cos 2x \, dx$$

can be determined by using the substitution $s = \sin 2x$. Hence we have $ds/dx = 2 \cos 2x$ and so the integral becomes

$$\int \tfrac{1}{8} \sin^2 2x \cos 2x \, dx = \int \tfrac{1}{16} s^2 \, ds = \tfrac{1}{48} s^3 + C$$

$$= \tfrac{1}{48} \sin^3 2x + C$$

Thus the solution is

$$\int \cos^4 x \sin^2 x \, dx = \tfrac{1}{8}x + \tfrac{1}{16} \sin 2x - \tfrac{1}{16}x - \tfrac{1}{32} \sin 4x - \tfrac{1}{16} \sin 2x$$

$$+ \tfrac{1}{48} \sin^3 2x + C$$

Example

Evaluate the integral

$$\int \sin^2 x \, dx$$

Since $\sin^2 x = \frac{1}{2}(1 - \cos 2x)$, see the Appendix, then the integral can be written as

$$\int \sin^2 x \, dx = \int \frac{1}{2}(1 - \cos 2x) \, dx$$

$$= \int \frac{1}{2} \, dx - \frac{1}{2} \int \cos 2x \, dx = \frac{x}{2} - \frac{1}{4} \sin 2x + C$$

There is no need for a substitution in this case.

Example

Evaluate the integral

$$\int \sin^4 x \cos x \, dx$$

Let $s = \sin x$. Then $ds/dx = \cos x$. Hence the integral becomes

$$\int \sin^4 x \cos x \, dx = \int s^4 \cos x \frac{ds}{\cos x} = \frac{s^5}{5} + C$$

$$= \frac{1}{5} \sin^5 x + C$$

Review problems

8 Evaluate, by using trigonometric relationships, the integrals:

(a) $\int \cos^2 x \, dx$, (b) $\int \sin 2x \cos x \, dx$, (c) $\int \tan^2 4x \, dx$

9 Evaluate, by making suitable substitutions, the integrals:

(a) $\int \frac{\sin x}{\cos^3 x} \, dx$, (b) $\int \sin^3 x \cos^4 x \, dx$, (c) $\int \frac{\cos^3 x}{\sin^2 x} \, dx$,

(d) $\int \cos^5 x \sin x \, dx$, (e) $\int \sin^2 2x \cos^3 2x \, dx$, (f) $\int \cos^4 x \sin^3 x \, dx$,

(g) $\int \sin^2 x \cos^4 x \, dx$

10 Using the substitution $u = \cos ax$, show that

$$\int \tan ax \, dx = \frac{1}{a} \ln\{\sec ax\} + C$$

11 Determine the area under the curve $y = \sin^2 x$ between $x = 0$ and $x = \pi$.

6.2.2 Trigonometric and hyperbolic substitutions

Many functions which involve the square root of a quadratic expression can be integrated by using a suitable trigonometric, or hyperbolic, substitution. Thus if a function contains $\sqrt{(a^2 - x^2)}$ then the substitution $x = a \sin \theta$ might be used. This changes the function into $\sqrt{[a^2(1 - \sin^2\theta)]} = a\cos\theta$ and permits integration using standard integrals. Table 6.1 lists trigonometric, hyperbolic and other substitutions that might be tried with different forms of functions.

Functions that are not of the form shown in the table can often be changed into such a form. For example, if we have the factor $\sqrt{(pz^2 + qz + r)}$ then completing the square enables us to write $\sqrt{[(pz + q/2)^2 + (r - q^2/4)]}$ and have a function of the form $\sqrt{(x^2 + a^2)}$, with $x = pz + q/2$ and $a^2 = r - q^2/4$. This is illustrated by an example in the following text.

Example

Evaluate the integral

$$\int \frac{1}{x^2 + 4} \, dx$$

This contains a factor of the form $x^2 + a^2 = [\sqrt{(x^2 + a^2)}]^2$ and thus we can try substitution 4 in table 6.1. Hence, if we let $x = 2 \tan \theta$, then $dx/d\theta = 2 \sec^2\theta$ and so

$$\int \frac{1}{x^2 + 4} \, dx = \int \frac{1}{4\tan^2\theta + 4} 2 \sec^2\theta \, d\theta$$

Since $\tan^2\theta + 1 = \sec^2\theta$ (see the Appendix), then

$$\int \frac{1}{x^2 + 4} \, dx = \int \frac{1}{4 \sec^2\theta} 2 \sec^2\theta \, d\theta$$

$$= \int \tfrac{1}{2} \, d\theta = \tfrac{1}{2}\theta + C$$

Since $x = 2 \tan \theta$ then $\theta = \tan^{-1}x/2$ and so

$$\int \frac{1}{x^2 + 4} \, dx = \tfrac{1}{2} \tan^{-1}\frac{x}{2} + C$$

Example

Evaluate the integral

$$\int \frac{1}{\sqrt{x^2+4}}\, dx$$

The above problem could be solved by using, as in the previous example, the substitution $x = 2 \tan \theta$ or by using the substitution $x = 2 \sinh \theta$. Then $dx/d\theta = \cosh \theta$ and so the integral becomes

$$\int \frac{1}{\sqrt{x^2+4}}\, dx = \int \frac{1}{\sqrt{4\sinh^2\theta+4}} \times \cosh \theta\, d\theta$$

Since $\sinh^2\theta + 1 = \cosh^2\theta$ (see the Appendix) the integral becomes

$$\int \frac{1}{\sqrt{x^2+4}}\, dx = \int \frac{1}{\sqrt{4\cosh^2\theta}} \times \cosh \theta\, d\theta = \int \tfrac{1}{2}\, d\theta$$

$$= \tfrac{1}{2}\theta + C$$

Since $x = 2 \sinh \theta$ then $\theta = \sinh^{-1} x/2$ and so

$$\int \frac{1}{\sqrt{x^2+4}}\, dx = \tfrac{1}{2}\sinh^{-1}\frac{x}{2} + C$$

If required the inverse sinh can be put in a logarithmic form (see the Appendix) to give

$$\int \frac{1}{\sqrt{x^2+4}}\, dx = \tfrac{1}{2}\ln\left(\frac{x}{2} + \sqrt{\frac{x^2}{4}+1}\right) + C$$

The problem could equally well have been solved using the substitution $x = 2 \tan \theta$. Then $dx/d\theta = 2 \sec^2\theta$ and so the integral becomes

$$\int \frac{1}{\sqrt{x^2+4}}\, dx = \int \frac{1}{\sqrt{4\tan^2\theta+4}} \times 2 \sec^2\theta\, d\theta$$

Since $\tan^2\theta + 1 = \sec^2\theta$ then

$$\int \frac{1}{\sqrt{x^2+4}}\, dx = \int \sec \theta\, d\theta = \ln(\operatorname{cosec} \theta - \cot \theta) + C$$

Since $x = 2 \tan \theta$ then $\theta = \tan^{-1}x/2$ and so

$$\int \frac{1}{\sqrt{x^2+4}}\, dx = \ln\left[\operatorname{cosec}\left(\tan^{-1}\frac{x}{2}\right) - \cot\left(\tan^{-1}\frac{x}{2}\right)\right] + C$$

The result is in a different form. However, it can be shown that the two solutions are identical.

Example

Evaluate the integral

$$\int \sqrt{4 - x^2}\, dx$$

This contains a factor of the form $\sqrt{(a^2 - x^2)}$ and thus is of the form indicated by item 3 in table 6.1. Hence, let $x = 2 \sin \theta$, i.e. $\sqrt{4} \sin \theta$. Then $dx/d\theta = 2 \cos \theta$ and so the integral becomes

$$\int \sqrt{4 - x^2}\, dx = \int \left(\sqrt{4 - 4\sin^2\theta} \right) 2 \cos \theta\, d\theta$$

$$= \int 4 \cos^2\theta\, d\theta$$

Since $\cos^2\theta = \frac{1}{2}(1 + \cos 2\theta)$ (see the Appendix) then

$$\int \sqrt{4 - x^2}\, dx = \int 2(1 + \cos 2\theta)\, d\theta$$

$$= \int 2\, d\theta + 2 \int \cos 2\theta\, d\theta$$

$$= 2\theta + \sin 2\theta + C$$

Since $\sin 2\theta = 2 \sin \theta \cos \theta$ (see the Appendix), then

$$\int \sqrt{4 - x^2}\, dx = 2\theta + 2 \sin \theta \cos \theta + C$$

Since $x = 2 \sin \theta$ then $\sin \theta = x/2$ and $\theta = \sin^{-1} x/2$. Also, since $\cos^2\theta + \sin^2\theta = 1$ then $\cos \theta = \sqrt{(1 - \sin^2\theta)} = \sqrt{(1 - x^2/4)}$. Thus

$$\int \sqrt{4 - x^2}\, dx = 2 \sin^{-1}\frac{x}{2} + 2\frac{x}{2} \times \sqrt{1 - \frac{x^2}{4}} + C$$

$$= 2 \sin^{-1}\frac{x}{2} + \frac{x}{2}\sqrt{4 - x^2} + C$$

This problem could equally well have been solved using the substitution $x = 2 \tanh \theta$ The answer would then be in an alternative form, but which can be shown to be the same solution as above.

Example

Evaluate the integral

$$\int \sqrt{x^2 + x - 6} \; dx$$

We can rewrite this integral as

$$\int \sqrt{x^2 + x - 6} \; dx = \int \sqrt{(x + \tfrac{1}{2})^2 - \tfrac{25}{4}} \; dx$$

With $u = x + \tfrac{1}{2}$ and $a^2 = 25/4$, we have an integral of $\sqrt{(u^2 - a^2)}$, with $du/dx = 1$. This is of the form given by item 6 in table 6.1. Thus, using the substitution $u = a \cosh \theta$ and $du/d\theta = a \sinh \theta$,

$$\int \sqrt{x^2 + x - 6} \; dx = \int \sqrt{u^2 - a^2} \; du$$

$$= \int \sqrt{a^2 \cosh^2 \theta - a^2} \; a \sinh \theta \; d\theta$$

But $\cosh^2\theta - \sinh^2\theta = 1$ (see the Appendix), hence

$$\int \sqrt{x^2 + x - 6} \; dx = \int a^2 \sinh^2\theta \; d\theta$$

Since $\cosh 2\theta = 1 + 2 \sinh^2\theta$ (see the Appendix),

$$\int \sqrt{x^2 + x - 6} \; dx = a^2 \int (\cosh 2\theta - 1) \; d\theta$$

$$= a^2 (\tfrac{1}{2} \sinh 2\theta - \theta) + C$$

Since $\sinh 2\theta = 2 \sinh \theta \cosh \theta$ (see the Appendix),

$$\int \sqrt{x^2 + x - 6} \; dx = a^2(\sinh \theta \cosh \theta - \theta) + C$$

But $\cosh^2\theta - \sinh^2\theta = 1$ (see the Appendix) and so we can write $\sinh \theta$ in terms of $\cosh \theta$ and obtain

$$\int \sqrt{x^2 + x - 6} \; dx = a^2 \left[\cosh \theta \sqrt{\cosh^2\theta - 1} - \theta \right] + C$$

$$= \frac{25}{8} \left[\frac{2u}{5} \sqrt{\left\{ \left(\frac{2u}{5} \right)^2 - 1 \right\}} - \cosh^{-1} \frac{2u}{5} \right] + C$$

$$= \tfrac{1}{4}(2x + 1) \sqrt{x^2 + x - 6}$$

$$- \tfrac{25}{8} \cosh^{-1} \left(\frac{2x + 1}{5} \right) + C$$

The inverse cosh can be put into logarithmic form (see the Appendix) to give

$$\int \sqrt{x^2 + x - 6} \, dx = \tfrac{1}{4}(2x + 1)\sqrt{x^2 + x - 6}$$

$$- \tfrac{25}{8} \ln\left(\frac{2x+1}{5} + \sqrt{\frac{(2x-1)^2}{25} - 1} \right) + C$$

Review problems

12 Evaluate, using trigonometric substitutions, the integrals:

(a) $\int \dfrac{1}{1+x^2} \, dx$, (b) $\int \dfrac{\sqrt{9-x^2}}{x^2} \, dx$, (c) $\int \dfrac{1}{9+x^2} \, dx$,

(d) $\int (9 - x^2) \, dx$, (e) $\int \dfrac{1}{\sqrt{9-x^2}} \, dx$, (f) $\int \dfrac{2}{4+x^2} \, dx$

13 Evaluate, using hyperbolic substitutions, the integrals:

(a) $\int \dfrac{x}{\sqrt{1+x^2}} \, dx$, (b) $\int \dfrac{2}{\sqrt{x^2-9}} \, dx$, (c) $\int \sqrt{x^2+4} \, dx$,

(d) $\int \dfrac{1}{x^2 \sqrt{1+x^2}} \, dx$, (e) $\int \sqrt{x^2 - 9} \, dx$

14 Evaluate the integrals:

(a) $\int \sqrt{x^2 + 4x - 5} \, dx$, (b) $\int \dfrac{1}{\sqrt{x^2 - 6x + 10}} \, dx$

6.2.3 The substitution tan x/2

A very useful substitution that is often used to put integrals into a suitable form for integration is $t = \tan x/2$. This gives

$$\frac{dt}{dx} = \tfrac{1}{2} \sec^2 \frac{x}{2}$$

But $\sec^2\theta = 1 + \tan^2\theta$, hence

$$\sec^2 \frac{x}{2} = 1 + \tan^2 \frac{x}{2} = 1 + t^2$$

Hence

$$\frac{dt}{dx} = \tfrac{1}{2}(1 + t^2)$$

and so when making this substitution we replace dx by

$$dx = \frac{2}{1+t^2} dt \qquad\qquad [5]$$

With $t = \tan x/2$ we can also develop relationships for tan x, sin x and cos x. Thus writing

$$\tan x = \tan\left[2\left(\frac{x}{2}\right)\right]$$

and then using the relationship tan $2\theta = (2 \tan \theta)/(1 - \tan^2\theta)$, see the Appendix,

$$\tan x = \frac{2\tan\frac{x}{2}}{1 - \tan^2\frac{x}{2}} = \frac{2t}{1 - t^2} \qquad\qquad [6]$$

We can represent this tangent of x in terms of the sides of a right-angled triangle with $2t$ for the opposite side of the right-angled triangle and $(1 - t^2)$ for the adjacent side (figure 6.1). The application of the Pythagoras theorem gives for the diagonal side $\sqrt{[(2t)^2 + (1 - t^2)^2]} = 1 + t^2$. Then

Fig. 6.1 Right-angled triangle

$$\sin x = \frac{2t}{1+t^2} \qquad\qquad [7]$$

$$\cos x = \frac{1-t^2}{1+t^2} \qquad\qquad [8]$$

Example

Evaluate the integral

$$\int \frac{1}{\sin x} dx$$

Let $t = \tan x/2$, then by using equations [7] and [5] the integral becomes

$$\int \frac{1}{\sin x} dx = \int \frac{1+t^2}{2t} \times \frac{2}{1+t^2} dt = \int \frac{1}{t} dt$$

and hence

$$\int \frac{1}{\sin x} dx = \ln t + C = \ln (\tan x/2) + C$$

Example

Evaluate the integral

$$\int \frac{1}{5 + 4\cos x}\, dx$$

Let $t = \tan x/2$, then using equation [8]

$$5 + 4\cos x = 5 + 4\left(\frac{1 - t^2}{1 + t^2}\right) = \frac{5 + 5t^2 + 4 - 4t^2}{1 + t^2} = \frac{9 + t^2}{1 + t^2}$$

and hence with equation [5] the integral becomes

$$\int \frac{1}{5 + 4\cos x}\, dx = \int \frac{1 + t^2}{9 + t^2} \times \frac{2}{1 + t^2}\, dt$$

$$= 2\int \frac{1}{9 + t^2}\, dt$$

We can evaluate this integral by making a second substitution. Let $t = 3\tan\theta$. Then $dt/d\theta = 3\sec^2\theta$ and so the integral becomes

$$2\int \frac{1}{9 + t^2}\, dt = 2\int \frac{1}{9 + 9\tan^2\theta} 3\sec^2\theta\, d\theta = 2\int \tfrac{1}{3}\, d\theta$$

and so

$$2\int \frac{1}{9 + t^2}\, dt = \tfrac{2}{3}\theta + C = \tfrac{2}{3}\tan^{-1}\tfrac{t}{3} + C$$

Hence

$$\int \frac{1}{5 + 4\cos x}\, dx = \tfrac{2}{3}\tan^{-1}\left(\tfrac{1}{3}\tan\tfrac{x}{2}\right) + C$$

Review problems

15 Evaluate the integrals:

(a) $\int \frac{1}{\cos x}\, dx$, (b) $\int \frac{1}{3 + 2\cos x}\, dx$

6.2.4 Change of limits with change of variable

Consider the integration of $\sqrt{(1 - x^2)}$ between the limits of $x = 0$ and $x = 1$, i.e.

$$\int_0^1 \sqrt{1 - x^2}\, dx$$

Ignore, for the moment, the limits. Using the substitution $x = \sin \theta$, with consequently $dx/d\theta = \cos \theta$, then

$$\int \sqrt{1-x^2} \, dx = \int \sqrt{1-\sin^2\theta} \times \cos\theta \, d\theta$$

$$= \int \cos^2\theta \, d\theta$$

$$= \int \tfrac{1}{2}(1+\cos 2\theta) \, d\theta$$

$$= \tfrac{1}{2}\theta + \tfrac{1}{4}\sin 2\theta + C$$

Since $\sin 2\theta = 2 \sin \theta \cos \theta$ (see the Appendix), then we can write

$$\int \sqrt{1-x^2} \, dx = \tfrac{1}{2}\theta + \tfrac{1}{2}\sin\theta \cos\theta + C$$

We have $x = \sin \theta$ and so $\theta = \sin^{-1} x$. Since $\cos^2\theta = 1 - \sin^2\theta$ then $\cos \theta = \sqrt{(1 - \sin^2\theta)}$ and so

$$\int \sqrt{1-x^2} \, dx = \tfrac{1}{2}\sin^{-1}x + \tfrac{1}{2}x\sqrt{1-x^2} + C$$

Now, if we consider the limits of $x = 0$ to $x = 1$ then we have

$$\int_0^1 \sqrt{1-x^2} \, dx = \left[\tfrac{1}{2}\sin^{-1}x + \tfrac{1}{2}x\sqrt{1-x^2} + C \right]_0^1$$

and so

$$\int_0^1 \sqrt{1-x^2} \, dx = \tfrac{\pi}{4} + 0 - 0 - 0 = \tfrac{\pi}{4}$$

There is a simpler way of obtaining this answer. That is to substitute the limits at the same time as substituting the function. Thus when we make the substitution $x = \sin \theta$ then when $x = 0$ we have a limit for θ given by $0 = \sin \theta$, i.e. $\theta = 0$. With $x = 1$ we have a limit for θ given by $1 = \sin \theta$, i.e. $\theta = \pi/2$. Thus substituting both the function and the limits we have

$$\int_0^1 \sqrt{1-x^2} \, dx = \int_0^{\pi/2} \sqrt{1-\sin^2\theta} \times \cos\theta \, d\theta$$

$$= \int_0^{\pi/2} \cos^2\theta \, d\theta$$

$$= \int_0^{\pi/2} \tfrac{1}{2}(1+\cos 2\theta) \, d\theta$$

$$= \left[\tfrac{1}{2}\theta + \tfrac{1}{4}\sin 2\theta + C \right]_0^{\pi/2}$$

$$= \tfrac{\pi}{4}$$

Review problems

16 Evaluate the following integrals:

(a) $\int_{-\infty}^{+\infty} \dfrac{x^2}{(4+x^2)}\,dx$ using $x = 2\tan\theta$ substitution,

(b) $\int_0^2 \dfrac{x}{\sqrt{1+2x^2}}\,dx$ using $u = \sqrt{1+2x^2}$ substitution,

(c) $\int_0^1 \sqrt{4-x^2}\,dx$ using $x = 2\sin\theta$ substitution

6.3 Integration by parts

The product rule for differentiation (see section 2.1.3) gives

$$\frac{d}{dx}(uv) = u\frac{dv}{dx} + v\frac{du}{dx}$$

where u and v are functions of x. Integrating both sides of this equation with respect to x gives

$$\int \frac{d}{dx}(uv)\,dx = \int u\frac{dv}{dx}\,dx + \int v\frac{du}{dx}\,dx$$

and so

$$uv = \int u\frac{dv}{dx}\,dx + \int v\frac{du}{dx}\,dx$$

This can be rearranged to give

$$\int u\frac{dv}{dx}\,dx = uv - \int v\frac{du}{dx}\,dx \qquad [9]$$

This is the equation for *integration by parts*. It is often written as

$$\int u\,dv = uv - \int v\,du$$

Integration by parts exchanges the integral on the left-hand side of the equation involving a product for a product of two functions and another integral. The procedure used is to choose one term of the product that is being integrated to be u and the other to be dv/dx. Then du/dx and v are determined with the hope that the resulting integral on the right-hand side of the equation is easier to evaluate than the one we started with. In choosing which part of the initial integral to be u and which dv/dx we might try letting dv/dx be the most complicated part, or letting u be the part whose derivative is a simpler function than u.

Example

Evaluate, using integration by parts, the integral

$$\int x e^x \, dx$$

Since e^x is the most complicated part of the product we will try it as dv/dx. Thus we let

$$u = x, \text{ with consequently } \frac{du}{dx} = 1$$

and

$$\frac{dv}{dx} = e^x, \text{ with consequently } v = \int \frac{dv}{dx} \, dx = \int e^x \, dx = e^x$$

Then equation [9] gives

$$\int u \frac{dv}{dx} \, dx = uv - \int v \frac{du}{dx} \, dx$$

and so

$$\int x e^x \, dx = x e^x - \int e^x \times 1 \, dx = x e^x - e^x + C$$

Note that there is no need to include a constant of integration when evaluating $v = \int e^x \, dx$, i.e. $v = e^x + A$. This is because when we use this value of v in the equation it cancels out, as the following shows.

$$\int x e^x \, dx = x(e^x + A) - \int (e^x + A) \times 1 \, dx$$

$$= x e^x + Ax - e^x - Ax + C$$

$$= x e^x - e^x + C$$

In this example, x was selected as u. Consider what would have happened if e^x had been selected as u. Let

$$u = e^x \text{ with consequently } \frac{du}{dx} = e^x$$

and

$$\frac{dv}{dx} = x \text{ with consequently } v = \int \frac{dv}{dx} \, dx = \int x \, dx = \frac{x^2}{2}$$

Equation [9] for integration by parts gives

$$\int u \frac{dv}{dx} dx = uv - \int v \frac{du}{dx} dx$$

and thus

$$\int x e^x dx = e^x \times \frac{x^2}{2} - \int \frac{x^2}{2} e^x dx$$

The new integral is further from being capable of being integrated than the one we started with. This plainly was the wrong choice of *u*.

Example

Evaluate, using integration by parts, the integral

$$\int x^2 \cos x \, dx$$

Since the differentiation of x^2 produces a simpler term we will try that as *u*. Thus we let

$$u = x^2, \text{ with consequently } \frac{du}{dx} = 2x$$

and

$$\frac{dv}{dx} = \cos x, \text{ with consequently } v = \int \frac{dv}{dx} dx = \int \cos x \, dx = \sin x$$

Hence integration by parts gives

$$\int u \frac{dv}{dx} dx = uv - \int v \frac{du}{dx} dx$$

and thus

$$\int x^2 \cos x \, dx = x^2 \sin x - \int \sin x \times 2x \, dx$$

This integral can be evaluated by a second use of integration by parts. Let

$$u = x, \text{ with consequently } \frac{du}{dx} = 1$$

and

$$\frac{dv}{dx} = \sin x, \text{ with consequently } v = \int \frac{dv}{dx} dx = \int \sin x \, dx = -\cos x$$

Hence, integration by parts gives

$$\int u \frac{dv}{dx}\,dx = uv - \int v \frac{du}{dx}\,dx$$

and so

$$\int x \sin x\,dx = x(-\cos x) - \int(-\cos x) \times 1\,dx$$

$$= -x \cos x + \sin x + C$$

Hence

$$\int x^2 \cos x\,dx = x^2 \sin x - x \cos x + \sin x + C$$

The above is an example of a method that is often referred to as *integration by successive reduction*. Many expressions can be integrated by stages, the integral obtained at each stage being simpler than the one at the preceding stage. Eventually an integral is reached which can be evaluated.

Example

Evaluate the integral

$$\int \ln x\,dx$$

Integration by parts involves the integration of an expression that is a product. The above integral can be put into the form of a product by writing it as

$$\int \ln x \times 1\,dx$$

For such an integral, let

$$u = \ln x, \text{ with consequently } \frac{du}{dx} = \frac{1}{x}$$

and

$$\frac{dv}{dx} = 1, \text{ with consequently } v = \int \frac{dv}{dx}\,dx = \int 1\,dx = x$$

Hence integration by parts gives

$$\int u \frac{dv}{dx}\,dx = uv - \int v \frac{du}{dx}\,dx$$

and

$$\int \ln x \times 1 \, dx = (\ln x)x - \int x \times \frac{1}{x} \, dx = x \ln x - x + C$$

Review problems

17 Evaluate, using integration by parts, the following integrals:

(a) $\int x \ln x$, (b) $\int x e^{2x} \, dx$, (c) $\int x^2 e^{-3x} \, dx$, (d) $\int x \sin x \, dx$,

(e) $\int \sec^3 x \, dx$, (f) $\int x^3 \cos x \, dx$, (g) $\int x \sec^2 x \, dx$,

(h) $\int 3x \sin 2x \, dx$, (i) $\int \ln 4x \, dx$

Hint: in (e) try the integral of $\sec^2 x \sec x$.

18 Evaluate, using integration by parts, the following integrals:

(a) $\int_0^1 x^2 e^x \, dx$, (b) $\int_0^{\pi/2} 3 e^x \cos 2x \, dx$, (c) $\int_0^1 x \ln x \, dx$

6.3.1 Integration of $e^{ax} \cos bx$

Oscillations with displacements from their rest positions described by either $\cos bx$ or $\sin bx$ are often damped in such a way that the resulting displacement is described by $e^{ax} \cos bx$ or $e^{ax} \sin bx$. Such oscillations might be those of an oscillating beam or a current in an electrical circuit.

Consider the integration of these functions, i.e. evaluate

$$\int e^{ax} \cos bx \, dx$$

Let

$$u = e^{ax} \text{ with consequently } \frac{du}{dx} = a e^{ax}$$

and

$$\frac{dv}{dx} = \cos bx \text{ with consequently}$$
$$v = \int \frac{dv}{dx} \, dx = \int \cos bx \, dx = \frac{1}{b} \sin x$$

Integration by parts gives

$$\int u \frac{dv}{dx} \, dx = uv - \int v \frac{du}{dx} \, dx$$

and so

$$\int e^{ax}\cos bx\,dx = \frac{e^{ax}\sin x}{b} - \int \frac{a}{b}e^{ax}\sin x\,dx + A \qquad [10]$$

Taking this second integral,

$$\int \frac{a}{b}e^{ax}\sin x\,dx = \frac{a}{b}\int e^{ax}\sin x\,dx$$

let

$$u = e^{ax} \text{ with consequently } \frac{du}{dx} = a\,e^{ax}$$

and

$$\frac{dv}{dx} = \sin x \text{ with consequently}$$

$$v = \int \frac{dv}{dx}\,dx = \int \sin bx\,dx = -\frac{1}{b}\cos bx$$

Integration by parts for this integral gives

$$\int u\frac{dv}{dx}\,dx = uv - \int v\frac{du}{dx}\,dx$$

and so

$$\int e^{ax}\sin x\,dx = -\frac{e^{ax}\cos bx}{b} + \int \frac{a}{b}e^{ax}\cos bx\,dx + B$$

Thus equation [10] becomes

$$\int e^{ax}\cos bx\,dx = \frac{e^{ax}\sin bx}{b} + \frac{a\,e^{ax}\cos bx}{b^2}$$

$$-\frac{a^2}{b^2}\int e^{ax}\cos bx\,dx + C$$

We have ended up, after using integration by parts twice, with the same integral on both sides of the equation. Thus we may combine the two integrals to give

$$\left(1 + \frac{a^2}{b^2}\right)\int e^{ax}\cos bx\,dx = \frac{e^{ax}(b\,\sin bx + a\cos bx)}{b^2} + C$$

and so

$$\int e^{ax}\cos bx\,dx = \frac{e^{ax}(b\,\sin bx + a\cos bx)}{a^2 + b^2} + C \qquad [11]$$

Similarly we can obtain

$$\int e^{ax}\sin bx\,dx = \frac{e^{ax}(b\,\sin bx - a\cos bx)}{a^2 + b^2} + C \qquad [12]$$

Example

Evaluate the integral

$$\int e^{2x}\sin 3x\,dx$$

We can integrate this by using the technique outlined above. The result is thus the form of equation given by equation [12]. Thus by substituting suitable values in equation [12] we have

$$\int e^{2x}\sin 3x\,dx = \frac{e^{2x}(3\,\sin 3x - 2\cos 3x)}{3^2 + 2^2} + C$$

$$= \frac{e^{2x}(3\,\sin 3x - 2\cos 3x)}{13} + C$$

Review problems

19 Evaluate the following integrals:

(a) $\int e^{2x}\cos x\,dx$, (b) $\int e^{x}\sin 2x\,dx$

20 The velocity of an object performing damped simple harmonic motion is given by the equation

$$v = e^{-at}\sin \omega t$$

Derive an equation for the distance travelled in an oscillation of half a cycle, i.e. from $t = 0$ to $t = \pi/\omega$.

5.4 Integration by partial fractions

Many integrals involving fractions may be integrated by expressing the integral as the sum of two or more simpler fractions, termed *partial fractions*, which can then be integrated individually. For example, the fraction

$$\frac{3x+4}{x^2+3x+2} = \frac{3x+4}{(x+1)(x+2)}$$

Table 6.2 Partial fractions

Denominator containing	Form of expression	Form of partial fractions
1 Linear factors	$\dfrac{f(x)}{(x+a)(x+b)(x+c)}$	$\dfrac{A}{(x+a)}+\dfrac{B}{(x+b)}+\dfrac{C}{(x+c)}$
2 Repeated linear factors	$\dfrac{f(x)}{(x+a)^3}$	$\dfrac{A}{(x+a)^3}+\dfrac{B}{(x+a)^2}+\dfrac{C}{(x+a)}$
3 Quadratic factor	$\dfrac{f(x)}{(ax^2+bx+c)(x+d)}$	$\dfrac{Ax+B}{(ax^2+bx+c)}+\dfrac{C}{(x+d)}$
4 Repeated quadratic factors	$\dfrac{f(x)}{(ax^2+bx+c)^2}$	$\dfrac{Ax+B}{(ax^2+bx+c)^2}+\dfrac{Cx+D}{(ax^2+bx+c)}$

can be expressed as the partial fractions

$$\frac{3x+4}{(x+1)(x+2)}=\frac{1}{x+1}+\frac{1}{x+2}$$

Thus

$$\int\frac{3x+4}{x^2+3x+2}\,dx=\int\frac{1}{x+1}\,dx+\int\frac{1}{x+2}\,dx$$

and then it is a comparatively simple matter to evaluate the integral.

The top line of a fraction is called the numerator and the bottom line the denominator. The degree of either the numerator or the denominator is the highest power of the variable in the expression. For example, the numerator $3x+4$ has the degree 1 while the denominator x^2+3x+2 has the degree 2. When the degree of the denominator is greater than that of the numerator then an expression can be resolved into partial fractions. For such situations, table 6.2 outlines the form that can be taken by the partial fractions for different types of denominator. The values of the constants A, B, C, etc. can be found by either making use of the fact that the equality between the fraction and its partial fractions must be true for all values of the variable x or that the coefficients of x^n in the fraction must equal those of x^n when the partial fractions are multiplied out.

To illustrate this, consider the example discussed earlier in this section. The partial fractions of

$$\frac{3x+4}{(x+1)(x+2)}$$

are given by item 1 in table 6.2 as

$$\frac{A}{x+1} + \frac{B}{x+2}$$

Then for the expressions to be equal we must have

$$\frac{3x+4}{(x+1)(x+2)} = \frac{A}{x+1} + \frac{B}{x+2} = \frac{A(x+2) + B(x+1)}{(x+1)(x+2)}$$

Thus

$$3x + 4 = A(x + 2) + B(x + 1)$$

Consider the requirement that this relationship is true for all values of x. Then, when $x = -1$ we must have

$$-3 + 4 = A(-1 + 2) + B(-1 + 1)$$

Hence $A = 1$. When $x = -2$ we must have

$$-6 + 4 = A(-2 + 2) + B(-2 + 1)$$

Hence $B = 1$. Alternatively, we could have determined these constants by multiplying out the expression and considering the coefficients, i.e.

$$3x + 4 = A(x + 2) + B(x + 1) = Ax + 2A + Bx + B$$

Thus, for the coefficients of x to be equal we must have

$$3 = A + B$$

and for the constants to be equal

$$4 = 2A + B$$

These two simultaneous equations can be solved to give A and B.

When the degree of the denominator is equal to or less than that of the numerator, the denominator must be divided into the numerator until the result is the sum of terms with the remainder fraction term having a denominator which is of higher degree than its numerator.

Consider, for example, the fraction

$$\frac{x^3 - x^2 - 3x + 1}{x^2 - 3x + 2}$$

The numerator has a degree of 3 and the denominator a degree of 2. Thus, dividing has to be used.

$$x^2 - 3x + 2 \overline{\smash{\big)}\, x^3 - x^2 - 3x + 1} \quad \overset{x + 2}{}$$

$$\begin{array}{r} \underline{x^3 - 3x^2 + 2x} \\ 2x^2 - 5x \\ \underline{2x^2 - 6x + 4} \\ x - 4 \end{array}$$

Thus

$$\frac{x^3 - x^2 - 3x + 1}{x^2 - 3x + 2} = x + 2 + \frac{x - 4}{x^2 - 3x + 2}$$

The fractional term can then be simplified using partial fractions.

$$\frac{x - 4}{x^2 - 3x + 2} = \frac{x - 4}{(x - 1)(x - 2)} = \frac{A}{x - 1} + \frac{B}{x - 2}$$

Hence we must have

$$x - 4 = A(x - 2) + B(x - 1)$$

When $x = 1$ we have $-3 = -A$ and so $A = 3$. When $x = 2$ we have $-2 = B$. Hence the partial fractions are $3/(x - 1)$ and $-1/(x - 2)$. Thus

$$\frac{x^3 - x^2 - 3x + 1}{x^2 - 3x + 2} = x + 2 + \frac{3}{x - 1} - \frac{1}{x - 2}$$

The procedure for obtaining partial fractions can thus be summarised as:

1 If the degree of the denominator is equal to, or less, than that of the numerator, divide the denominator into the numerator to obtain the sum of a polynomial plus a fraction which has the degree of the denominator greater than that of the numerator.
2 Write the denominator in the form of linear factors, i.e. of the form $(ax + b)$, or irreducible quadratic factors, i.e. of the form $(ax^2 + bx + c)$ and which cannot be factored further.
3 Write the fraction as a sum of partial fractions (see table 6.2 for possible forms).
4 Determine the unknown constants which occur with the partial fractions by equating the fraction with the partial fractions and either solving the equation for specific values of x or equating the coefficients of equal powers of x.
5 Replace the constants in the partial fractions with their values.

Example

Use partial fractions to simplify the following expression:

$$\frac{2x-19}{(x-2)^2(x+3)}$$

The denominator is of higher degree than the numerator, so it can be transformed into partial fractions without first dividing. The denominator contains a repeated factor and so the partial fraction form is

$$\frac{2x-19}{(x-2)^2(x+3)} = \frac{A}{(x-2)^2} + \frac{B}{(x-2)} + \frac{C}{(x+3)}$$

Thus we must have

$$2x - 19 = A(x+3) + B(x-2)(x+3) + C(x-2)^2$$

When $x = -3$ then we have $-6 - 19 = 25C$ and so $C = -1$. When $x = 2$ then $4 - 19 = 5A$ and so $A = -3$. When $x = 0$ then we have $-19 = 3A - 6B + 4C$ and so $B = 1$. Thus

$$\frac{2x-19}{(x-2)^2(x+3)} = -\frac{3}{(x-2)^2} + \frac{1}{(x-2)} - \frac{1}{(x+3)}$$

Example

Use partial fractions to simplify the following expression:

$$\frac{3x}{(x^2 - 2x + 5)(x+1)}$$

This expression contains a quadratic and thus

$$\frac{3x}{(x^2 - 2x + 5)(x+1)} = \frac{A+Bx}{(x^2 - 2x + 5)} + \frac{C}{(x+3)}$$

Hence we must have

$$3x = (A + Bx)(x + 3) + C(x^2 - 2x + 5)$$

With $x = -3$ then we have $-9 = C(9 + 6 + 5)$ and so $C = -9/20$. With $x = 0$ we have $0 = 3A + 5C$ and so $A = 3/4$. With $x = 1$ we have $3 = 4(A + B) + 4C$ and so $B = 9/20$. Thus the partial fractions are

$$\frac{3x}{(x^2 - 2x + 5)(x+1)} = \frac{15+9x}{20(x^2 - 2x + 5)} - \frac{9}{20(x+3)}$$

Example

Evaluate the integral

$$\int \frac{8(x+1)}{x(x^2-4)}\,dx$$

Partial fractions can be used to simplify the integral. Thus

$$\frac{8(x+1)}{x(x^2-4)} = \frac{8(x+1)}{x(x-2)(x+2)} = \frac{A}{x} + \frac{B}{x-2} + \frac{C}{x+2}$$

Hence

$$8x + 8 = A(x-2)(x+2) + Bx(x+2) + Cx(x-2)$$

When $x = 2$ we have $24 = 8B$ and so $B = 3$. When $x = -2$ we have $-8 = 8C$ and so $C = -1$. When $x = 0$ we have $8 = -4A$ and so $A = -2$. Thus

$$\frac{8(x+1)}{x(x^2-4)} = -\frac{2}{x} + \frac{3}{x-2} - \frac{1}{x+2}$$

Thus we now require the integral

$$-\int \frac{2}{x}\,dx + \int \frac{3}{x-2}\,dx - \int \frac{1}{x+2}\,dx$$

For these integrals we have

$$-\int \frac{2}{x}\,dx = -2\ln x + A$$

If we let $u = x - 2$ then $du/dx = 1$ and so we have

$$\int \frac{3}{x-2}\,dx = \int \frac{3}{u}\,du = 3\ln u + B = 3\ln(x-2) + B$$

If we let $u = x + 2$ then $du/dx = 1$ and so we have

$$-\int \frac{1}{x+2}\,dx = -\int \frac{1}{u}\,du = -\ln u + C = -\ln(x+2) + C$$

Hence

$$\int \frac{8(x+1)}{x(x^2-4)}\,dx = -2\ln x + 3\ln(x-2) - \ln(x+2) + D$$

$$= \ln\left(\frac{(x-2)^3}{x^2(x+2)}\right) + D$$

Example

Evaluate the integral

$$\int \frac{x^2}{(x^2+1)^2}\,dx$$

We can write this fraction, which is a repeated quadratic, as

$$\frac{x^2}{(x^2+1)^2} = \frac{Ax+B}{x^2+1} + \frac{Cx+D}{(x^2+1)^2}$$

Hence we must have

$$x^2 = (Ax+B)(x^2+1) + Cx + D$$

$$= Ax^3 + Bx^2 + Ax + B + Cx + D$$

Equating coefficients gives: for those of x^3 we have $A = 0$, for those of x^2 we have $B = 1$, for those of x we have $0 = A + C$ and so $C = 0$, for those of constants we have $0 = B + D$ and so $D = -1$. Hence we have

$$\int \frac{x^2}{(x^2+1)^2}\,dx = \int \frac{1}{x^2+1}\,dx - \int \frac{1}{(x^2+1)^2}\,dx$$

The first integral is the same form as the standard integral given as item 14 in table 5.1 and gives $\tan^{-1}x + C$. The second integral we can evaluate by means of a substitution. Let $x = \tan\theta$. This then gives $dx/d\theta = \sec^2\theta$ and so

$$\int \frac{1}{(x^2+1)^2}\,dx = \int \frac{1}{(\tan^2\theta+1)^2}\sec^2\theta\,d\theta$$

Since $\tan^2\theta + 1 = \sec^2\theta$ then

$$\int \frac{1}{(x^2+1)^2}\,dx = \int \cos^2\theta\,d\theta = \int \tfrac{1}{2}(1+\cos 2\theta)\,d\theta$$

$$= \frac{\theta}{2} + \tfrac{1}{4}\sin 2\theta + C = \frac{\theta}{2} + \tfrac{1}{2}\sin\theta\cos\theta + C$$

Since $\tan\theta = x$ we can represent θ as the angle in the right-angled triangle shown in figure 6.2. Then we have

$$\int \frac{1}{(x^2+1)^2}\,dx = \frac{\theta}{2} + \frac{1}{2}\left(\frac{x}{\sqrt{x^2+1}}\right)\left(\frac{1}{\sqrt{x^2+1}}\right) + C$$

$$= \tfrac{1}{2}\tan^{-1}x + \frac{x}{2(x^2+1)} + C$$

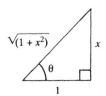

$\sqrt{(1+x^2)}$

x

θ

1

Fig. 6.2 Example

Thus

$$\int \frac{x^2}{(x^2+1)^2}\, dx = \tan^{-1}x - \tfrac{1}{2}\tan^{-1}x - \frac{x}{2(x^2+1)} + C$$

$$= \tfrac{1}{2}\tan^{-1}x - \frac{x}{2(x^2+1)} + C$$

Review problems

21 Use partial fractions to simplify the following expressions:

(a) $\dfrac{x-6}{(x-1)(x-2)}$, (b) $\dfrac{x+5}{x^2+3x+2}$, (c) $\dfrac{x-13}{(x-1)(x+3)}$,

(d) $\dfrac{x^2}{(x-2)^2(x-1)}$, (e) $\dfrac{5x-1}{(x+1)^2(x-2)}$, (f) $\dfrac{x+5}{(x-1)(x^2+x+1)}$,

(g) $\dfrac{x^2+2}{(x+4)(x-2)}$, (h) $\dfrac{3x^2-x-2}{x+2}$, (i) $\dfrac{x^3+x+6}{(x-2)(x-4)}$

22 Evaluate, using partial fractions, the following integrals:

(a) $\displaystyle\int \frac{10x-8}{x^2-8x+12}\, dx$, (b) $\displaystyle\int \frac{x-8}{(x+1)(x-2)}\, dx$, (c) $\displaystyle\int \frac{x^3}{x-2}\, dx$,

(d) $\displaystyle\int \frac{1}{x^2-9}\, dx$, (e) $\displaystyle\int \frac{2x^3+3x^2-3}{2x^2-x-1}\, dx$,

(f) $\displaystyle\int \frac{5x^2-7x-7}{(x-1)(x+2)(2x+1)}\, dx$, (g) $\displaystyle\int \frac{1}{(x+1)^2(x^2+4)}\, dx$,

(h) $\displaystyle\int \frac{3x+2}{x(x^2+4)(x-1)}\, dx$

23 The work done w in moving an object from a distance of 2 m to 4 m is given by

$$w = \int_2^4 \frac{8}{x(x^2+4)}\, dx$$

Determine the work done.

Further problems

24 Determine the following integrals:

(a) $\displaystyle\int 2x^5\, dx$, (b) $\displaystyle\int 4\sin 2x\, dx$, (c) $\displaystyle\int \frac{5}{x}\, dx$, (d) $\displaystyle\int (x^2+4x-2)\, dx$,

(e) $\displaystyle\int (2+e^{-5x})\, dx$, (f) $\displaystyle\int (2\sin x - 4\cos x)\, dx$, (g) $\displaystyle\int 2\tan^2 x\, dx$,

(h) $\int \cos 5x \sin 2x\, dx$, (i) $\int \sin 3x \sin 5x\, dx$, (j) $\int (1-x)^2\, dx$,

(k) $\int (x-1)^{1/2}\, dx$, (l) $\int x e^{x^2}\, dx$, (m) $\int (1+x)^{0.2}\, dx$,

(n) $\int \dfrac{1}{(1+x)^2}\, dx$, (o) $\int 6(3x-1)^4\, dx$, (p) $\int (2x+1)(x^2+x)\, dx$,

(q) $\int \dfrac{1}{3+\sqrt{x}}\, dx$, (r) $\int \dfrac{x}{\sqrt{1-x}}\, dx$, (s) $\int \dfrac{2}{\sqrt{4-x^2}}\, dx$,

(t) $\int \sqrt{4-x^2}\, dx$, (u) $\int \dfrac{1}{(x^2+1)^2}\, dx$, (v) $\int \dfrac{x}{\sqrt{4-x^2}}\, dx$,

(w) $\int \dfrac{1}{2\cos x+3}\, dx$, (x) $\int \dfrac{1}{2\sin x+1}\, dx$, (y) $\int \dfrac{1}{5\sin x+13}\, dx$,

(z) $\int \dfrac{1}{2\sin x+3}\, dx$, (aa) $\int x^2 \ln x\, dx$, (ab) $\int x^2 e^{2x}\, dx$,

(ac) $\int x e^{3x}\, dx$, (ad) $\int 2x \sin 2x\, dx$, (ae) $\int \dfrac{\ln x}{x}\, dx$,

(af) $\int 4x^2 \sin 2x\, dx$, (ag) $\int x^4 \ln x\, dx$, (ah) $\int \sqrt{x}\, \ln x\, dx$,

(ai) $\int e^{3x}\cos 2x\, dx$, (aj) $\int e^{2x}\sin 5x\, dx$, (ak) $\int \dfrac{2}{(x+1)(x-1)}\, dx$,

(al) $\int \dfrac{x}{x^2-1}\, dx$, (am) $\int \dfrac{x^2}{x^2-4}\, dx$, (an) $\int \dfrac{x^2}{(x-1)(2x+3)}\, dx$,

(ao) $\int \dfrac{4x-7}{x^2-4x+2}\, dx$, (ap) $\int \dfrac{x+5}{x^2+6x+9}\, dx$,

(aq) $\int \dfrac{x+3}{(x+2)(x+5)}\, dx$, (ar) $\int \dfrac{x+1}{(x+2)(x-1)}\, dx$

25 The velocity v of an object as a function of time t is given by the equation

$$v = 4t^2 + 2t$$

Determine the equation relating the displacement and time.

26 Determine the function which has a graph with a gradient given by

$$\frac{dy}{dx} = x^2 + 2x + 1$$

27 Using the substitution $u = \sin ax$, show that

$$\int \cot ax\, dx = \frac{1}{a}\ln(\sin ax) + C$$

28 The velocity v of an object is given by the equation

$$v = 10(1 + t)^{3/2}$$

Derive an equation for the displacement x as a function of the time t.

29 Determine the values of the following integrals:

(a) $\int_0^1 \frac{x}{(x^2 + 2)^2} \, dx$ using the substitution $u = x^2 + 2$,

(b) $\int_0^5 \frac{1}{\sqrt{25 - x^2}} \, dx$ using the substitution $x = 5 \sin \theta$,

(c) $\int_{-2}^2 \frac{\sqrt{x+2}}{x+6} \, dx$ using the substitution $u = \sqrt{x+2}$,

(d) $\int_0^1 \sqrt{1 + x^2} \, dx$ using the substitution $x = \tan \theta$,

(e) $\int_1^2 x e^{-x^2} \, dx$ using the substitution $u = x^2$,

(f) $\int_0^3 \sqrt{9 - x^2} \, dx$ using the substitution $x = 3 \sin \theta$,

(g) $\int_0^{\pi/2} x \cos 2x \, dx$ using integration by parts,

(h) $\int_0^1 x e^{3x} \, dx$ using integration by parts,

(i) $\int_0^\infty e^{-2x} \sin x \, dx$ using integration by parts,

(j) $\int_1^2 \frac{2x - 1}{(x + 1)^2} \, dx$ using partial fractions,

(k) $\int_4^5 \frac{1}{x^2 - 4} \, dx$ using partial fractions,

(l) $\int_3^5 \frac{x^2 + 1}{(x + 3)(x - 2)} \, dx$ using partial fractions

30 The electric field strength E at the centre of a semicircular arc of wire of radius r when it is carrying a charge q is given by

$$E = \frac{q}{r} \int_{-\pi/2}^{\pi/2} \theta \cos \theta \, d\theta$$

Determine the electric field strength.

31 What is the area under a graph of y against x between $x = 1$ and $x = 3$ if $y = \ln x$?

32 What is the area under a graph of y against t between $t = 0$ and $t = \pi/\omega$, i.e. a half cycle, if $y = e^{\alpha t}\sin \omega t$?

33 The time t taken for the concentration x in a chemical reaction to reach the value X is given by

$$t = \int_0^X \frac{1}{k(a-x)(b-x)}\, dx$$

where a, b and k are constants. Evaluate the integral.

7 Areas and volumes

7.1 Areas by integration

This chapter is a consideration of how integration can be applied to the determination of areas and volumes. These might be, for example, the area under a graph or perhaps the surface area and volume of a solid.

This section is about the areas of flat surfaces, typically the areas under graphs. The area between a graph of a function $f(x)$ plotted against x and the x-axis when considered between the limits of $x = a$ and $x = b$, as in figure 7.1(a), can be obtained by dividing it into a number of equal width narrow vertical strips and then summing the areas of all the strips between $x = a$ and $x = b$. Thus if the width of a strip is δx then the area of a strip $\delta A = y\, \delta x$. The sum of all the strips is

$$\text{area} = \sum_{x=a}^{x=b} y\, \delta x$$

In the limit when δx tends to 0 we have the definite integral

$$\text{area} = \int_a^b y\, dx \qquad [1]$$

We could have considered the area between the graph line and the y-axis, as in figure 7.1(b). Then we would have used horizontal strips and the area would have been

$$\text{area} = \int_a^b x\, dy \qquad [2]$$

See section 5.1.2 for a further discussion of the definite integral.

Thus suppose, for example, we have a function $y = x^2$ and require the area under the graph of this function between the ordinates at $x = 1$ and $x = 2$. The area is then given by equation [1] as

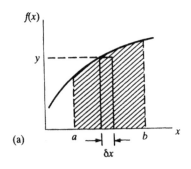

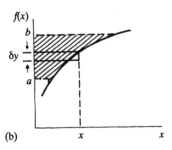

Fig. 7.1 Area under a curve

142

$$\text{area} = \int_1^2 x^2 \, dx = \left[\tfrac{1}{3}x^3 + C\right]_1^2 = 1 \text{ square unit}$$

In finding an area by integration we are finding the sum of the areas of strips, i.e. terms like $y \, \delta x$. The width of the strip δx is always a positive increment but y can be positive or negative, i.e. the strip is above or below the x-axis. When y is negative then the area of the strip is negative, when positive then the area is positive. Thus areas above the x-axis are positive and below the x-axis are negative. To illustrate this, consider the graph of the function $y = x(x - 2)$. Figure 7.2 shows the graph. For an area between the ordinates $x = 0$ and $x = 2$ we have

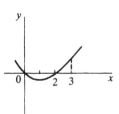

Fig. 7.2 $y = x(x - 2)$

$$\text{area} = \int_0^2 x(x - 2) \, dx = \left[\tfrac{1}{3}x^3 - x^2 + C\right]_0^2 = -\tfrac{4}{3} \text{ square units}$$

The integral gives a negative area because the area is below the x-axis. For an area between $x = 2$ and $x = 3$ we have

$$\text{area} = \int_2^3 x(x - 2) \, dx = \left[\tfrac{1}{3}x^3 - x^2 + C\right]_2^3 = +\tfrac{4}{3} \text{ square units}$$

The integral gives a positive area because the area is above the x-axis. For an area between $x = 0$ and $x = 3$ we have

$$\text{area} = \int_0^3 x(x - 2) \, dx = \left[\tfrac{1}{3}x^3 - x^2 + C\right]_0^3 = 0$$

This is because the area above the x-axis is cancelled by the area below the axis.

Thus if we require the total area enclosed when there are positive and negative areas we must find separately the positive and negative areas so that we can take the sum of them, disregarding the signs, to give the actual area.

Example

The work done in stretching a strip of material is the area under its force–distance graph. If, for a particular material, the force F in newtons is given by $F = 100x$, with x in metres, determine the work done in stretching the material from an extension $x = 0$ to $x = 0.1$ m.

The work done is the area under the graph between these limits, as illustrated by figure 7.3, and is given by

$$W = \int_0^{0.1} 100x \, dx = [50x^2 + C]_0^{0.1} = 0.5 \text{ J}$$

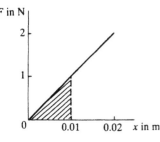

Fig. 7.3 Example

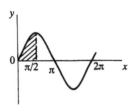

Fig. 7.4 Example

Example

Determine the area between the graph of $y = \sin x$ and the x-axis between the limits (a) 0 and $\pi/2$, (b) 0 and π, (c) π and 2π, (d) 0 and 2π.

(a) Figure 7.4 shows the graph and the required area. It is given by

$$\text{area} = \int_0^{\pi/2} \sin x \, dx = [-\cos x + C]_0^{\pi/2} = 1 \text{ square unit}$$

(b) The area is given by

$$\text{area} = \int_0^{\pi} \sin x \, dx = [-\cos x + C]_0^{\pi} = 2 \text{ square units}$$

(c) The area is given by

$$\text{area} = \int_{\pi}^{2\pi} \sin x \, dx = [-\cos x + C]_{\pi}^{2\pi} = -2 \text{ square units}$$

(d) The area is given by

$$\text{area} = \int_0^{2\pi} \sin x \, dx = [\cos x + C]_0^{2\pi} = 0$$

Example

The pressure p of an ideal gas is related to its volume V by Boyle's law, i.e. $pV = $ a constant. A particular gas has a volume of 3 m^3 at a pressure of 140 kPa. The work done when a gas expands from a volume V_1 to V_2 is given by

$$\text{work done} = \int_{V_1}^{V_2} p \, dV$$

Hence determine the work done when the gas expands from 2 m^3 to 4 m^3.

For the gas we have $pV = k$. The integral can thus be written as

$$\text{work done} = \int_{V_1}^{V_2} p \, dV = \int_{V_1}^{V_2} \frac{k}{V} \, dV = [k \ln V + C]_{V_1}^{V_2}$$

Hence, since $pV = 140 \times 3 = 420$ kPa m^3 (or kJ) we have

$$\text{work done} = 420(\ln 4 - \ln 2) = 291.1 \text{ kJ}$$

Example

Determine the equation for the area of a circle.

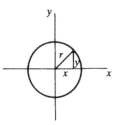

Fig. 7.5 Example

Applying the Pythagoras theorem to the circle shown in figure 7.5 gives the equation for a circle as

$$x^2 + y^2 = r^2$$

Now consider the area of the positive quadrant of the circle. We are only considering the positive quadrant, rather than the full enclosed area, since negative and positive areas are included and we do not want them to cancel out. Indeed, it is obvious that since there will be an equal amount of negative and positive area that integrating over the full area will give a zero answer. Thus we plan to find the area of each quadrant independently and then sum the areas, disregarding signs, to give the total area. The area of the positive quadrant is given by

$$\text{area} = \int_0^r y \, dx = \int_0^r \sqrt{r^2 - x^2} \, dx$$

We can solve this integration by substitution. Let $x = r \sin \theta$. Then $dx/d\theta = r \cos \theta$. When $x = 0$ then $\theta = 0$ and when $x = r$ we have $\theta = \pi/2$. Then

$$\int_0^r \sqrt{r^2 - x^2} \, dx = \int_0^{\pi/2} \sqrt{r^2 - r^2 \sin^2 \theta} \times r \cos \theta \, d\theta$$

$$= \int_0^{\pi/2} r^2 \cos^2 \theta \, d\theta = \int_0^{\pi/2} \frac{r^2}{2}(1 + \cos 2\theta) \, d\theta$$

$$= \frac{r^2}{2}\left[\theta - \tfrac{1}{2} \sin 2\theta\right]_0^{\pi/2}$$

$$= \frac{\pi r^2}{4}$$

The total area of the circle is thus πr^2.

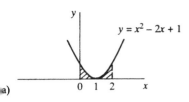

(a)

$y = x^2 - 2x + 1$

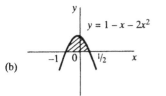

(b)

$y = 1 - x - 2x^2$

Review problems

1 Determine the area between the function $y = 2x^2 + 3$ and the x-axis and enclosed by the ordinates $x = 1$ and $x = 2$.

2 Determine the area between the function $y = x^2 - 3x + 2$ and the x-axis and enclosed by the ordinates (a) $x = 0$ and $x = 1$, (b) $x = 1$ and $x = 2$, (c) $x = 0$ and $x = 2$.

3 Determine the area between the function $y = \cos x$ and the x-axis and enclosed by the ordinates at (a) $x = 0$ and $x = \pi/2$, (b) $\pi/2$ and π, (c) 0 and π.

4 Determine the marked areas indicated in figure 7.6.

5 The velocity v, in metres per second, of an object is given by $v = 4 + 3t^2$, where t is the time in seconds. The distance moved by the body in a time interval is the area under the velocity–

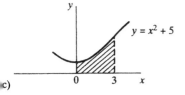

(c)

$y = x^2 + 5$

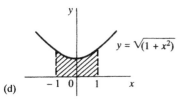

(d)

$y = \sqrt{(1 + x^2)}$

Fig. 7.6 Problem 4

time graph in that time. Determine the distance moved by the object between $t = 0$ and $t = 3$ s.

6   Determine the area between the function $y = 4 \sin 2x$ and the x-axis and enclosed by the ordinates $x = 0$ and $x = \pi/3$.

7   Determine the area between the function $y = 2/(x + 1)$ and the x-axis and enclosed by the ordinates $x = 1$ and $x = 5$.

8   Determine the area of an ellipse given that the equation of an ellipse is

$$\frac{x^2}{a^2} + \frac{y^2}{b^2} = 1$$

For an ellipse centred on the origin, a is the value of x where the ellipse cuts the x-axis and b the value of y where the ellipse cuts the y-axis.

7.1.1 Area between curves

Consider the determination of the area between the two curves $y = f(x)$ and $y = g(x)$ shown in figure 7.7(a). The area between the function $y = f(x)$ and the x-axis and between the ordinates of the points of intersection, a and b, of the two functions is

$$\int_a^b f(x)\, dx$$

The area between the function $y = g(x)$ and the x-axis and between the same ordinates a and b is

$$\int_a^b g(x)\, dx$$

The required area is the difference between these two areas, namely

$$\text{area} = \int_a^b f(x)\, dx - \int_a^b g(x)\, dx = \int_a^b [f(x) - g(x)]\, dx \qquad [3]$$

An alternative way of arriving at the above equation is to divide the area between the two curves into vertical strips of equal width. A strip will have an area of $[f(x) - g(x)]\, \delta x$. Thus the total area between the points of intersection a and b will be the sum of all these strips, i.e.

$$\text{area} = \sum_{x=a}^{x=b} [f(x) - g(x)]\, \delta x$$

In the limit as δx tends to 0 we have equation [3].

The area given by an integral will be positive is $f(x)$ is greater

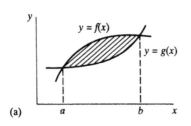

(a)

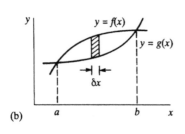

(b)

Fig. 7.7 Area between two curves

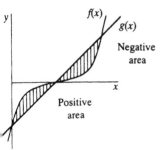

Fig. 7.8 Positive and negative areas

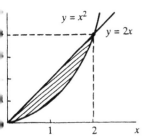

Fig. 7.9 Example

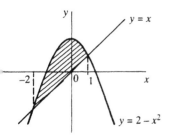

Fig. 7.10 Example

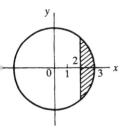

Fig. 7.11 Problem 10

than $g(x)$ for all points between a and b, and negative if otherwise. Thus if the curves describing the two functions cross between a and b there will be positive and negative areas. Figure 7.8 shows such a situation. To obtain the true area between the curves it is necessary to carry out the integration for the positive and negative area parts separately and then, disregarding the signs of the areas, add the results.

Example

Determine the area bounded by the curves $y = 2x$ and $y = x^2$.

Figure 7.9 shows the graph. The points of intersection are when the coordinates of the two curves are equal, i.e. the values of x and y that simultaneously satisfy both equations. For identical values of y we must have $2x = x^2$. This requires $x = 0$ or $x = 2$. When $x = 0$ we must have $y = 0$. When $x = 2$ we must have $y = 4$. Thus the required area is

$$\int_0^2 [2x - x^2]\, dx = \left[x^2 - \tfrac{1}{3}x^3 + C\right]_0^2 = \tfrac{4}{3} \text{ square units}$$

Example

Determine the area bounded by the curves $y = 2 - x^2$ and $y = x$.

The points of intersection between the two functions are when the coordinates of the two curves are the same. This is when we have $y = x = 2 - x^2$. This quadratic of $x^2 + x - 2 = 0$ can be written as $(x + 2)(x - 1) = 0$ and we have $x = -2$ and $x = 1$. The coordinates of the points of intersection are thus $(-2, -2)$ and $(1, 1)$. Figure 7.10 shows the graph. The required area is thus

$$\int_{-2}^1 [(2 - x^2) - x]\, dx = \left[2x - \tfrac{1}{3}x^3 - \tfrac{1}{2}x^2 + C\right]_{-2}^1 = \tfrac{9}{2} \text{ square units}$$

Review problems

9 Determine the areas bounded by the curves:
   (a) $y = 2x - 3x^2$ and $y = 0$,
   (b) $y = x^2$ and $y = x^3$,
   (c) $y = x^3$ and $y = 2x$,
   (d) $y = x^2 + 2x + 1$ and $y = 3x + 5$
   Note: it is a useful aid to seeing the form of the problem to sketch the graphs roughly.
10 Determine the shaded area indicated in figure 7.11.

7.2 Volumes by integration

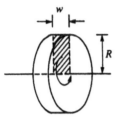

Fig. 7.12 Generating a disc

This section is concerned with the determination of the volumes of what are termed *solids of revolution*. Such solids have circular cross-sections and can be considered to be generated by the rotation of an area about an axis. Figure 7.12 illustrates this, showing how a disc can be generated by rotating an area. The volume is $\pi R^2 w$, where R is the radius of the disc and w its width. We can generate a wineglass shape by rotating the area between the graph of $y = x^2$, a y-ordinate and the y-axis about the y-axis, as illustrated in figure 7.13.

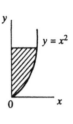

Fig. 7.13 Generating a wine glass shaped volume

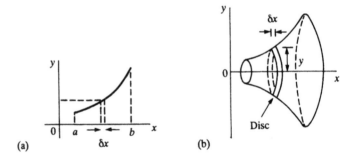

Fig. 7.14 (a) Area rotated, (b) the solid generated

We can consider any solid of revolution to be made up of discs. Figure 7.14 illustrates this when the solid is formed by revolution of areas round the x-axis. The volume of the solid then becomes the sum of the volumes of these discs. The volume of a disc of radius y and width δx is $\pi y^2\, \delta x$. The volume of the solid is thus the sum of all such discs between $x = a$ and $x = b$, i.e.

$$\text{volume} = \sum_{x=a}^{x=b} \pi y^2\, \delta x$$

In the limit as δx tends to 0 we have

$$\text{volume} = \int_a^b \pi y^2\, dx \qquad [4$$

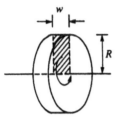

Fig. 7.15 Rotation about the y-axis

If the solid of revolution had been formed by a revolution of areas round the y-axis (figure 7.15) then the volume of each disc radius x and width δy, is $\pi x^2 \, \delta y$. The volume of the solid is then

$$\text{volume} = \sum_{y=a}^{y=b} \pi x^2 \, \delta y$$

In the limit as δy tends to 0 we have

$$\text{volume} = \int_a^b \pi x^2 \, dy \qquad [5]$$

Example

Determine the volume generated by rotating the area between the function $y = e^x$ and the x-axis, and between the ordinates $x = 1$ and $x = 2$, about the x-axis.

Figure 7.16 shows the area being rotated. Using equation [4], then

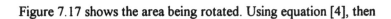

$$\text{volume} = \int_a^b \pi y^2 \, dx = \int_1^2 \pi (e^x)^2 \, dx = \left[\frac{\pi}{2} e^{2x} + C \right]_1^2$$

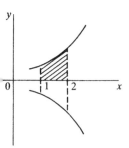

Fig. 7.16 Example

$$= \frac{\pi}{2}(e^4 - e^2) = 74.16 \text{ cubic units}$$

Example

Determine the volume generated by rotating the area between the function $y = 2x$ and the x-axis, and between the ordinates $x = 1$ and $x = 4$, about the x-axis.

Figure 7.17 shows the area being rotated. Using equation [4], then

$$\text{volume} = \int_a^b \pi y^2 \, dx = \int_1^4 \pi (2x)^2 \, dx = \left[\frac{4\pi}{3} x^3 + C \right]_1^4$$

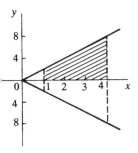

Fig. 7.17 Example

$$= \frac{4\pi}{3}(4^3 - 1^3) = 263.9 \text{ cubic units}$$

Review problems

11 Determine the volumes generated by rotating the areas between the following functions and the x-axis about the x-axis:

(a) $y = \sqrt{\sin x}$ between $x = 0$ and $x = \pi$,
(b) $y = x^2 + 4$ between $x = 1$ and $x = 4$,
(c) $y = x^3$ between $x = 0$ and $x = 3$,
(d) $y = 3 + 2 \sin x$ between $x = 0$ and $x = \pi$

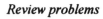

12 Determine the volumes generated by rotating the area enclosed in the positive quadrant by $y = 9 - x^2$ about (a) the x-axis, (b) the y-axis.

7.2.1 Volume generated by areas between curves

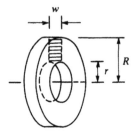

Fig. 7.18 Generating a perforated disc

Consider a disc containing a hole, as illustrated in figure 7.18. The volume of this disc is the volume of a disc of radius R minus the volume of a disc of radius r, i.e. $\pi(R^2 - r^2)w$. It is formed by rotating the marked area about the x-axis. We can consider other volumes that are formed by rotating an area between two curves as the sum of a number of such discs. Thus if we rotate the area between the two functions $y = f(x)$ and $y = g(x)$, shown in figure 7.19, about the x-axis then the volume generated is the volume generated by rotating $f(x)$, between the limits a and b, about the x-axis minus the volume generated by rotating $g(x)$, between the same limits, about the x-axis, i.e.

$$\text{volume} = \int_a^b \pi\{f(x)\}^2 \, dx - \int_a^b \pi\{g(x)\}^2 \, dx$$

$$= \int_a^b \pi[\{f(x)\}^2 - \{g(x)\}^2] \, dx \qquad [6]$$

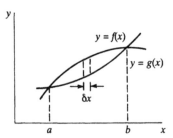

Fig. 7.19 Generating a perforated volume

Example

Determine the volume generated by rotating the area bounded by the functions $y = \sqrt{x}$ and $y = x^2$ about the x-axis.

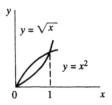

Fig. 7.20 Example

Figure 7.20 shows graphs of the two functions and the area concerned. The boundaries of the area will be when $\sqrt{x} = x^2$, i.e. when $x = 0$ or 1. Hence, using equation [6] we have

$$\text{volume} = \int_0^1 \pi[\{\sqrt{x}\}^2 - \{x^2\}^2] \, dx$$

$$= \int_0^1 \pi[x - x^4] \, dx = \pi\left[\tfrac{1}{2}x^2 - \tfrac{1}{5}x^5\right]_0^1$$

$$= \frac{3\pi}{10} \text{ cubic units}$$

Review problems

13 Determine the volumes generated by rotating the area bounded by the following functions about the x-axis:
(a) $y = x^2/4$ and $y = 2x$,
(b) $y = 4\sqrt{x}$ and $y = x^2$,
(c) $y = x^2$ and $y = x^3$
14 What is the volume of material required for a sphere of radius 5 cm with a hole passing through its centre of radius 2 cm. Hint: the equation for a sphere is $x^2 + y^2 = r^2$.

7.3 Lengths of curves

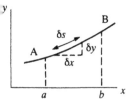

Fig. 7.21 Length of a curve

Consider the problem of determining the length of a curve. This might be, for example, the length of a cable suspended between two points and in the form of a catenary. For the graph shown in figure 7.21, if the length of a small segment of the curve is δs then the length of the curve between A and B is the sum of all such elements between A and B, i.e.

$$\text{length between A and B} = \sum_{x=a}^{x=b} \delta s$$

In the limit as δs tends to 0 we have

$$\text{length between A and B} = \int_a^b ds$$

But, for a small enough length we can write, using the Pythagoras theorem,

$$(\delta s)^2 = (\delta x)^2 + (\delta y)^2$$

This can be rewritten as

$$\left(\frac{\delta s}{\delta x}\right)^2 = 1 + \left(\frac{\delta y}{\delta x}\right)^2$$

$$\frac{\delta s}{\delta x} = \sqrt{1 + \left(\frac{\delta y}{\delta x}\right)^2} \qquad [7]$$

Thus, as $\delta s \to 0$, we have for the length between A and B

$$\text{length} = \int_a^b ds = \int_a^b \frac{ds}{dx}\,dx$$

$$= \int_a^b \sqrt{1 + \left(\frac{dy}{dx}\right)^2}\,dx \qquad [8]$$

Example

Determine the length of the curve $y = x^{3/2}$ between $x = 0$ and $x = 8$.

For $y = x^{3/2}$ we have $dy/dx = \frac{3}{2}x^{1/2}$ and so, using equation [8], the length of the curve is

$$\text{length} = \int_0^8 \sqrt{1 + \left(\frac{3}{2}x^{1/2}\right)^2}\,dx = \int_0^8 \frac{1}{2}\sqrt{4 + 9x}\,dx$$

We can evaluate this integral by substitution. Let $u = 4 + 9x$. Then

$du/dx = 9$ and, with $u = 76$ when $x = 8$ and $u = 4$ when $x = 0$, we have

$$\int_0^8 \tfrac{1}{2}\sqrt{4+9x}\ dx = \int_4^{76} \tfrac{1}{18}u^{1/2}\ du = \tfrac{1}{18}\left[\tfrac{2}{3}u^{3/2} + C\right]_4^{76}$$

and so the length $= 24.2$ units.

Review problems

15 Determine the length of the circumference of a circle, equation $x^2 + y^2 = r^2$.
   Hint: determine the length of the curve in a quadrant and then multiply the result by 4.
16 Determine the lengths of the curves given by the graphs of the following functions:
   (a) $y = \tfrac{1}{4}x^2$ from $x = 0$ to $x = 2$,
   (b) $y = \ln(1 - x^2)$ from $x = 0$ to $x = 1/4$,
   (c) $y = \ln(\cos x)$ from $x = 0$ to $x = \pi/4$
17 An electric cable hanging freely between two pylons sags and has the shape of a catenary. The catenary is defined by the equation $y = a\ \cosh(x/a)$. Determine the length of wire used if the pylons are a distance $2d$ apart.

7.4 Areas of surfaces of solids

To determine the area of the curved surface of a solid of revolution consider a basic element in the form of a disc. The curved surface of a disc is its circumference multiplied by the disc width. The curved surface area of a disc of radius y and width δs is thus given by $2\pi y\ \delta s$. Thus the total curved surface area of the solid generated by the revolution of the marked area in figure 7.22 is the sum of the curved surface areas of all the discs between the limits of $x = a$ and $x = b$. Thus

$$\text{curved surface area} = \sum_{x=a}^{x=b} 2\pi y\ \delta s$$

As δs tends to 0 then

$$\text{curved surface area} = \int_a^b 2\pi y\ ds$$

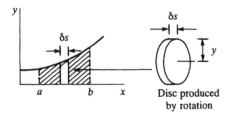

Fig. 7.22 Estimation of surface area

This can be written as

$$\text{curved surface area} = \int_a^b 2\pi y \frac{ds}{dx} \, dx$$

Hence, using equation [7] we can write

$$\text{curved surface area} = \int_a^b 2\pi y \sqrt{1 + \left(\frac{\delta y}{\delta x}\right)^2} \, dx \qquad [9]$$

Example

Determine the curved surface area of the surface formed between $x = 0$ and $x = 1$ by rotating the graph of $y = x^3$ about the x-axis.

Figure 7.23 shows the form of the graph and the solid produced by its rotation. Since $dy/dx = 3x^2$ then equation [8] gives

$$\text{area} = \int_0^1 2\pi x^3 \sqrt{1 + (3x^2)^2} \, dx = 2\pi \int_0^1 x^3 \sqrt{1 + 9x^4} \, dx$$

We can evaluate this integral by substitution. Let $u = 1 + 9x^4$. Then $du/dx = 36x^3$ and when $x = 1$ we have $u = 10$ and when $x = 0$ we have $u = 1$. The integral can thus be written as

$$\text{area} = 2\pi \int_1^{10} x^3 u^{1/2} \frac{du}{36x^3} = \frac{\pi}{18} \int_1^{10} u^{1/2} \, du$$

and so

$$\text{area} = \frac{\pi}{18} \left[\tfrac{2}{3} u^{3/2} + C \right]_1^{10} = 3.56 \text{ square units}$$

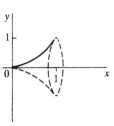

Fig. 7.23 Example

Review problems

18 Determine the areas of the curved surfaces of the solids generated by the rotation of the graphs of the following functions around the x-axis:
(a) $y = \sin x$ between $x = 0$ and $x = \pi$,
(b) $y = x^2/2$ between $x = 0$ and $x = 2$,
(c) $y^2 = x^3$ between $x = 0$ and $x = 4$

19 Determine the surface area of a mirror with a parabolic surface which can be considered to have been generated by the rotation of the parabola $y^2 = 4x$ between $x = 0$ and $x = 2$ about the x-axis.

20 Determine the surface area of a sphere.
Hint: determine the surface area of a quadrant of a circle when rotated, the equation of a circle being $x^2 + y^2 = r^2$, and then double it to obtain the required surface area.

Further problems

21 Determine the area between the function $y = 4x^3$ and the x-axis and enclosed by the ordinates $x = 2$ and $x = 5$.

22 Determine the area between the function $y = 3x^2 + 5$ and the x-axis and enclosed by the ordinates $x = 1$ and $x = 2$.

23 Determine the area between the function $y = x^3 - x$ and the x-axis and enclosed by the ordinates $x = -1$ and $x = +1$.

24 Determine the area between the function $y = \ln x$ and the x-axis and enclosed by the ordinates $x = 1$ and $x = 5$.

25 Determine the area between the function $y = 2(x + e^x)$ and the x-axis and enclosed by the ordinates $x = 0$ and $x = 2$.

26 The velocity v, in metres/second, of a moving object is related to the time travelled t, in seconds, by the equation $v = 2 + 3t^2$. Determine the distance travelled in the time interval from a time $t = 2$ s to $t = 5$s.

27 Determine the areas bounded by the curves:
   (a) $y = 6x - x^2$ and $y = x + 4$,
   (b) $y = x^2 - 4x + 1$ and $y = -x^2 + 2x + 1$,
   (c) $y = x^2 + 1$ and $y = 3 - x$,
   (d) $y = x^3 - x$, $y = 0$

28 For a circle of radius 4 units, determine the area cut off from the circle by a chord whose distance from the centre is 3 units.

29 Determine the volumes generated by rotating the areas between the following functions and the x-axis about the x-axis:
   (a) $y = 5x$ between $x = 1$ and $x = 4$,
   (b) $y = \sec x$ between $x = 0$ and $x = \pi/4$,
   (c) $y = x^3$ between $x = 0$ and $x = 1$,
   (d) $y = e^{2x}$ between $x = 1$ and $x = 2$,
   (e) $y = 5$ between $x = 0$ and $x = 2$

30 Determine the volumes generated by rotating the areas between the following functions and the y-axis about the y-axis:
   (a) $y = x^2$ between $y = 1$ and $y = 3$,
   (b) $y = \sqrt{x}$ between $y = 0$ and $y = 1$,
   (c) $y = 1/x$ between $y = 1$ and $y = 3$

31 Determine the volumes generated by rotating the area bounded by the following functions about the x-axis:
   (a) $y = 2\sqrt{x}$ and $y = x^2/2$,
   (b) $y = \sqrt{(25 - x^2)}$ and $y = 3$

32 Determine the lengths of the curves described by the graphs of the following functions:
   (a) $y^2 = 12x^3$ between $x = 0$ and $x = 1$,
   (b) $y^2 = 4x$ between $x = 0$ and $x = 2$,
   (c) $y^2 = x^3$ between $x = 0$ and $x = 5$,
   (d) $y = e^x$ from $x = 3/4$ to $x = 4/3$

33 Determine the areas of the surfaces of the solid generated by the rotation of the graphs of the following functions around the x-axis:

(a) $y^2 = \frac{1}{2}x^3$ between $x = 0$ and $x = 8$,

(b) $y = \frac{1}{4}x^2$ between $x = 0$ and $x = 4$,

(c) $y = \frac{3}{4}x$ between $x = 0$ and $x = 3$,

(d) $y = \frac{1}{3}x^3$ between $x = 0$ and $x = 3$

34 Determine the surface area of part of a sphere, the part being generated by the rotation of a circle of radius 3 units between $x = 0$ and $x = 2$ units about the x-axis.

35 A right circular cone of base radius r and height h can be considered to be generated by the rotation of the function described by $y = rx/h$, between the limits $x = 0$ and $x = h$, about the x-axis. Determine the surface area of the cone.

8 Moments

8.1 Moments

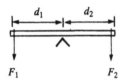

Fig. 8.1 The seesaw

There are many situations in mechanics where a summation of an infinite number of infinitesimally small products is required. Such situations can be handled by integration. Moments are one such quantity, this chapter giving illustrations.

The product of a force F and its perpendicular distance d from some axis is termed its *moment* about that axis.

$$\text{Moment of a force} = Fd \qquad [1]$$

Thus for a seesaw (figure 8.1) to balance, i.e. not to rotate, then the moment producing an anticlockwise rotation must balance the moment producing a clockwise rotation. If we consider the anti-clockwise moment to be positive and the clockwise moment to be negative, then this statement amounts to stating that the algebraic sum of the moments must be zero when the system is balanced, i.e.

$$\sum \text{moments} = 0$$

When any object is in equilibrium under the action of a number of parallel forces then the sum of the moments of the forces about some axis must be zero.

The above discussion refers to the moments of forces, the moment being the product of a force and a distance. Such moments are important in the consideration of the equilibrium of bodies. There are, however, other forms of moment which are used in engineering. The analysis of the stresses developed in the bending of beams depends on the shape and area of the cross-section of a beam. Such analysis involves the consideration of a beam in terms of it being made up of a large number of small elements of area at varying distances from an axis and then the summation to give the overall effect. Such analysis results in the summation of the products of area A and its distance d from some axis and is referred to as the *first moment of area*.

moment of area $= Ad$ [2]

In connection with the rotation of a body we have the *moment of inertia*. This can be termed a second moment of force since it involves the sum of the products of a large number of elements of the mass and the square of their distances from some axis.

The bending of beams can involve a *second moment of area*, this being the sum of the products of a large number of small elements of area and the square of their distances from some axis.

This chapter is about moments, section 8.2 being concerned with the moments arising from the weight or mass of a body and section 8.3 with the first moment of area. Second moments are considered in sections 8.4 and 8.5, the moment of inertia in section 8.4 and the second moment of area in section 8.5.

8.2 Centre of gravity

Consider a flat rectangular sheet of a material. It can be balanced in a horizontal position when supported at a particular point directly called the *centre of gravity*. The location of the centre of gravity of an object is important, for not only does it determine the position at which it can be supported and balance but also how it will behave when subject to forces. Thus, for example, the distance of the centre of gravity of a car above the ground will determine the angle at which the car can be tilted before it overturns.

We can think of the sheet as being made up of a large number of particles, each having weight. Suppose there to be a particle with a weight δw_1 at a distance of y_1 from some axis (figure 8.2), another with a weight δw_2 at a distance y_2, another with a weight δw_3 at a distance y_3, etc. Now if we take moments about that axis we will have

total moment $= \delta w_1 y_1 + \delta w_2 y_2 + \delta w_3 y_3 + \ldots$

If a single weight W is to replace all these forces then it must have the same moment, i.e.

$$W\bar{y} = \delta w_1 y_1 + \delta w_2 y_2 + \delta w_3 y_3 + \ldots = \sum \delta w\, y$$

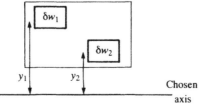

Fig. 8.2 Moments due to small elements of weight

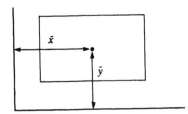

Fig. 8.3 The centre of gravity

where $\bar{y}$ is the distance of W from the axis. The centre of gravity is thus a distance $\bar{y}$ from the axis about which we have chosen to consider the moments (figure 8.3).

$$\bar{y} = \frac{\sum \delta w\, y}{W} \qquad [3]$$

In a similar way we can obtain, when considering an axis at right angles to the above axis,

$$\bar{x} = \frac{\sum \delta w\, x}{W} \qquad [4]$$

The centre of gravity has the coordinates $(\bar{x}, \bar{y})$. To determine the coordinate relating to the depth of the centre of gravity within the sheet we would need to take moments about a third axis which is mutually at right angles to the two previous axes. We would then obtain

$$\bar{z} = \frac{\sum \delta w\, z}{W} \qquad [5]$$

and so the coordinates $(\bar{x}, \bar{y}, \bar{z})$. For many items just the $\bar{x}$ and $\bar{y}$ coordinates suffice since they are of constant thickness and the $\bar{z}$ coordinate can be assumed to be just halfway through the thickness.

The weight of an object is the product of its mass m and the acceleration due to gravity g, i.e. mg. Thus, since the acceleration due to gravity appears in both the numerator and denominator of the above equations for the position of the centre of gravity, we can replace the equations by ones of the form

$$\bar{y} = \frac{\sum \delta m\, y}{M} \qquad [6]$$

and talk of the *centre of mass* of the body. The centre of gravity and the centre of mass refer to the same point in a body, provided the acceleration due to gravity is the same for all points in the body.

If a solid is symmetric about an axis then the centre of gravity or centre of mass must lie on that axis. Only then will the moments about that axis be zero. Thus for a sphere the centre of gravity lies at its centre, the solid being symmetric about axes through the centre. For a solid that can be generated by revolution then the centre of gravity must lie on the axis about which the revolution occurs. Thus when determining the position of the centre of gravity of solids of revolution it is generally simplest to take the axis of revolution as a coordinate axis.

Example

Determine the distance of the centre of gravity from one end of a uniform cross-section rod of uniform density.

The rod is symmetric about an axis at right angles to its length and through its centre. Thus the centre of gravity must lie on this axis and so a distance of half the length of the rod from one end.

Example

Determine the distance of the centre of gravity from one end of a uniform cross-section rod if the density of the rod varies as the square of the distance from the end.

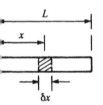

Fig. 8.4 Example

Consider an element of the rod of width δx a distance x from one end (figure 8.4). The volume of the element is $A\,\delta x$. Thus, if the density is ρ then the mass is $\rho A\,\delta x$. Since the mass varies as the square of x then the weight of the element can be written as $kx^2\,\delta x$, where k is a constant. The total weight of the rod is thus the sum of the weights of all these elements. If we let $\delta x \to 0$ then

$$\text{total weight} = \int_0^L kx^2\,dx = \tfrac{1}{3}kL^3$$

The moment of the element about the end is $(kx^2\,\delta x)x$. Thus

$$\text{total moment} = \int_0^L kx^3\,dx = \tfrac{1}{4}kL^4$$

The centre of gravity is thus given by equation [4] as at a distance $\bar{x}$ from that end where

$$\bar{x} = \frac{\tfrac{1}{4}kL^4}{\tfrac{1}{3}kL^3} = \tfrac{3}{4}L$$

Example

Determine the position of the centre of gravity for a solid hemisphere of constant density.

In this problem we will consider the location of the centre of mass, it being the same point as the centre of gravity. Because we are referring to an object of uniform density ρ then

$$\bar{x} = \frac{\sum \delta m\,x}{M} = \frac{\sum \rho \delta v\,x}{\rho V} = \frac{\sum \delta v\,x}{V} \qquad [7]$$

where v is the volume of an element and V the total volume of the solid.

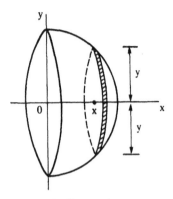

Fig. 8.5 Example

If we rotate a semicircle about its diameter then we obtain a solid hemisphere, as illustrated in figure 8.5. If we take the diameter as being on the x-axis then the centre of gravity will lie on the x-axis. The equation of a circle is $x^2 + y^2 = r^2$. Consider the disc segment shown in the figure. It has a volume of $\delta v = \pi y^2 \, \delta x$. The total volume V of the hemisphere is the sum of all the discs between $x = 0$ and $x = r$, i.e.

$$V = \sum_{x=0}^{x=r} \pi y^2 \, dx$$

Thus, in the limit when $\delta x \to 0$, we have

$$V = \int_0^r \pi y^2 \, dx$$

The moment of the disc about the y-axis is $(\pi y^2 \, \delta x)x$. Thus the sum of all such moments is

$$\text{moment} = \sum_{x=0}^{x=r} \pi y^2 x \, \delta x$$

Thus, in the limit when $\delta x \to 0$, we have

$$\text{moment} = \int_0^r \pi y^2 x \, dx$$

Thus the distance of the centre of mass from the y-axis is given by equation [7] as

$$\bar{x} = \frac{\int_0^r \pi y^2 x \, dx}{\int_0^r \pi y^2 \, dx}$$

Since $x^2 + y^2 = r^2$ then

$$\bar{x} = \frac{\int_0^r \pi(r^2 - x^2)x \, dx}{\int_0^r \pi(r^2 - x^2) \, dx}$$

We can solve the integral in the numerator by a substitution. Let $u = r^2 - x^2$. Then $du/dx = -2x$ and when $x = r$ we have $u = 0$ and when $x = 0$ we have $u = r^2$. Hence

$$\int_0^r \pi(r^2 - x^2)x \, dx = \int_{r^2}^0 \pi ux\left(-\frac{du}{2x}\right) = -\frac{\pi}{2}\int_{r^2}^0 u \, du$$

$$= -\frac{\pi}{2}\left[\tfrac{1}{2}u^2 + C\right]_{r^2}^0 = \tfrac{1}{4}\pi r^4$$

For the integral in the denominator we have

$$\int_0^r \pi(r^2 - x^2)\, dx = \pi\left[r^2 x - \tfrac{1}{3}x^3 + C \right]_0^r = \tfrac{2}{3}\pi r^3$$

Hence

$$\bar{x} = \frac{\tfrac{1}{4}\pi r^4}{\tfrac{2}{3}\pi r^3} = \tfrac{3}{8}r$$

Review problems

1   Determine the distance of the centre of gravity from one end of a uniform cross-section rod if the density of the rod varies as the cube of the distance from the end.
2   Determine the distance of the centre of gravity from the lighter end of a uniform cross-section rod if the mass per unit length of the rod a distance x from the lighter end is $(2 + x^2/2)$ kg/m and the rod is 4 m long.
3   Determine the centre of mass of a paraboloid generated by the rotation of the curve $y = x^2$, between $y = 0$ and $y = 4$, about the y-axis.

8.2.1 Centre of gravity of composite bodies

The term *composite body* is used to describe a body which can be considered to be made up of two or more simple shapes, such as spheres, cylinders, cubes, etc. The centre of gravity of a uniform density sphere lies at its centre, of a uniform density cylinder half way along its length and on the central axis, of a uniform density cube at its centre. Since we can treat an object as having all its weight acting at its centre of gravity then we can determine the centre of gravity of a composite body by using the weights and positions of the centres of gravity of each component part and determine the point of balance, i.e. the point about which the moments of the weights are zero. The following example illustrates the procedure.

Note that if there is a hole is an object, the hole can be regarded as having a negative weight acting at its centre of gravity.

Example

Determine the position of the centre of gravity of a mallet. The mallet has a rubber head in the form of a uniform density cylinder of diameter 80 mm and mass 750 g. The handle is 350 mm long and has a mass of 500 g.

Figure 8.6 shows the form of the mallet. Because both the handle and the head are symmetric about an axis down the centre of the

Fig. 8.6 Example

handle then the centre of gravity must lie along that axis. The centre of gravity of the handle will be midway along its central axis. The centre of gravity of the head will be at its centre. We can thus consider the entire weight of the handle to be acting at a distance from 175 mm from one end and the entire weight of the head to act at a distance of 40 mm from the handle. Taking moments about the handle end of the mallet gives

$$\text{total moment} = 500g \times 175 + 750g \times (40 + 350)$$

where g is the acceleration due to gravity. If all the mass of the mallet was concentrated at one point, the centre of gravity, then we would have to have

$$(500 + 750)g \times \bar{x} = 500g \times 175 + 750g \times (40 + 350)$$

Thus $\bar{x} = 304$ mm.

Review problems

4   Determine the position of the centre of gravity for the composite object shown in figure 8.7. It can be considered to be made up of a solid, uniform density, cylinder of diameter 60 mm and length 140 mm and a solid right circular cone of uniform density, diameter 60 mm and height 280 mm. The centre of gravity of a solid right circular cone lies in its central axis at one quarter of its height above the base.

5   Determine the position of the centre of gravity for the composite object shown in figure 8.8. It is a cube of side 400 mm with a cube of side 200 mm centrally removed from one side. The material has a uniform density.

Fig. 8.7 Problem 4

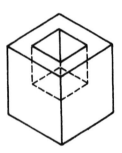

Fig. 8.8 Problem 5

8.3 Centroids

Consider an item of a constant thickness and constant density. The weight per unit area w is constant. Now consider an element of weight $\delta w_1 = w \delta A_1$, where δA_1 is a small element of area. The total moment is thus

$$\sum \delta w \, y = w \sum \delta A \, y$$

But the total weight is wA, where A is the total area. Hence

$$\bar{y} = \frac{w \sum \delta A \, y}{wA} = \frac{\sum \delta A \, y}{A} \tag{8}$$

Similarly

$$\bar{x} = \frac{\sum \delta A\, x}{A} \qquad [9]$$

The product of an element of area and its distance from some axis is called the *first moment of area*. The coordinates $(\bar{x}, \bar{y})$ give the location of the *centroid*.

The above discussion was concerned with the weight of a sheet of material. There are, however, other situations in engineering, e.g. in the bending of beams, where the first moment of area is involved and we can then consider the centroid as being the point in the plane of the area at which, when any axis is drawn through it, the first moment of the area about that axis is zero.

If an area is symmetric about an axis the centroid must lie on that axis, only then will the algebraic sum of the moments about the centroid be zero. If an area has two axes of symmetry then the position of the centroid must lie at the intersection of the axes of symmetry. Thus for a rectangle we can consider the area to be symmetrical about its two diagonals. Thus the centroid must lie at the point where the two diagonals intersect.

Consider the determination of the centroid for an area which is defined between a curve on a graph and the x-axis and the ordinates $x = a$ and $x = b$ (figure 8.9). We will divide the area into strips, each of width δx. The strip area is $\delta A = y\, \delta x$. The total area under the graph is thus

$$A = \sum_{x=a}^{x=b} y\, \delta x$$

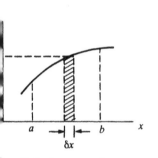

Fig. 8.9 Area under a graph

The moment about the y-axis of the area of a strip, i.e. its first moment of area, is $x\, \delta A = xy\, \delta x$. Hence the sum of all the moments of the areas of such strips is

$$\sum_{x=a}^{x=b} xy\, \delta x$$

Thus the centroid is, using equation [9], a distance of

$$\bar{x} = \frac{\displaystyle\sum_{x=a}^{x=b} xy\, \delta x}{\displaystyle\sum_{x=a}^{x=b} y\, \delta x}$$

from the y-axis. As δx tends to 0 then we obtain

$$\bar{x} = \frac{\displaystyle\int_{a}^{b} xy\, dx}{\displaystyle\int_{a}^{b} y\, dx} \qquad [10]$$

Now consider the position of the centroid from the x-axis. The strip defined in figure 8.9 is a rectangle and will have its centroid at its midpoint which is a distance $\frac{1}{2}y$ from the x-axis. We can consider all the area to be effectively located at this distance from the axis. Thus the first moment of area for the strip is

$$\tfrac{1}{2}y\,\delta A = \tfrac{1}{2}y(y\,\delta x) = \tfrac{1}{2}y^2\,\delta x$$

The sum of all these moments for the area under the graph is

$$\sum_{x=a}^{x=b} \tfrac{1}{2}y^2\,\delta x$$

Hence the distance of the centroid of the area from the x-axis is

$$\bar{y} = \frac{\displaystyle\sum_{x=a}^{x=b} \tfrac{1}{2}y^2\,\delta x}{\displaystyle\sum_{x=a}^{x=b} y\,\delta x}$$

When $\delta x \to 0$ then

$$\bar{y} = \frac{\tfrac{1}{2}\int_a^b y^2\,dx}{\int_a^b y\,dx} \qquad\qquad [11]$$

We can similarly consider the position of the centroid for an area defined between a curve on a graph and the y-axis and the ordinates $y = a$ and $y = b$ (figure 8.10). The area of a strip is $x\,\delta y$. Thus the total area between the graph and the y-axis is

$$A = \sum_{x=a}^{x=b} x\,\delta y$$

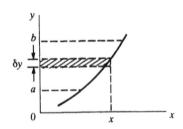

Fig. 8.10 Area under a graph

The moment of the area for a strip about the x-axis is $y\,\delta A = xy\,\delta y$. Hence the sum of the moments about the x-axis of all the strips under the graph is

$$\sum_{x=a}^{x=b} xy\,\delta y$$

Thus the distance of the centroid from the x-axis is

$$\bar{y} = \frac{\displaystyle\sum_{x=a}^{x=b} xy\,\delta y}{\displaystyle\sum_{x=a}^{x=b} x\,\delta y}$$

In the limit as $\delta y \to 0$ then

$$\bar{y} = \frac{\int_a^b xy\,dy}{\int_a^b x\,dy} \tag{12}$$

Now consider the position of the centroid from the y-axis. The strip defined in figure 8.10 is a rectangle and will have its centroid at its midpoint which is a distance $\frac{1}{2}x$ from the y-axis. We can consider all the area to be effectively located at this distance from the axis. Thus the first moment of area for the strip is

$$\tfrac{1}{2}x\,\delta A = \tfrac{1}{2}x(x\,\delta y) = \tfrac{1}{2}x^2\,\delta y$$

Thus the sum of the moments of all the strips under the curve is

$$\sum_{x=a}^{x=b} \tfrac{1}{2}x^2\,\delta y$$

Thus the distance of the centroid from the y-axis is

$$\bar{x} = \frac{\displaystyle\sum_{x=a}^{x=b} \tfrac{1}{2}x^2\,\delta y}{\displaystyle\sum_{x=a}^{x=b} x\,\delta y}$$

In the limit as $\delta y \to 0$ then

$$\bar{x} = \frac{\tfrac{1}{2}\int_a^b x^2\,dy}{\int_a^b x\,dy} \tag{13}$$

We can also define a moment of area for a line. A line can be considered as an area which has a constant width. Thus for a line of length s, when we consider an element of that length of δs then if we consider it to have a width w we have an area of $w\,\delta s$. The moment of this area about the x-axis is $yw\,\delta s$. Hence, in the limit as $\delta s \to 0$,

$$\bar{y} = \frac{\int_0^s yw\,ds}{\int_0^s w\,ds}$$

$$= \frac{\int_0^s y\,ds}{s} \tag{14}$$

Similarly we can derive

$$\bar{x} = \frac{\int_0^s x \, ds}{s}$$ [15]

Example

Determine the position of the centroid of a sheet of uniform density and constant thickness and which has an area bounded by the graph of the function $y = 4x - x^2$ and the x-axis.

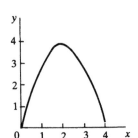

Fig. 8.11 Example

Figure 8.11 shows the graph of the function. When $y = 0$ then we have $4x = x^2$ and so $x = 0$ or 4. Considering the moments about the x-axis, then using equation [10]

$$\bar{x} = \frac{\int_a^b xy \, dx}{\int_a^b y \, dx} = \frac{\int_0^4 x(4x - x^2) \, dx}{\int_0^4 (4x - x^2) \, dx} = \frac{\left[\frac{4}{3}x^3 - \frac{1}{4}x^4 + C\right]_0^4}{\left[2x^2 - \frac{1}{3}x^3 + C\right]_0^4} = 2$$

This is what we should have expected since the shape is symmetrical about the $x = 2$ axis. Using equation [11] we have

$$\bar{y} = \frac{\frac{1}{2}\int_a^b y^2 \, dx}{\int_a^b y \, dx} = \frac{\frac{1}{2}\int_0^4 (4x - x^2)^2 \, dx}{\int_0^4 (4x - x^2) \, dx} = \frac{\frac{1}{2}\int_0^4 (16x^2 - 8x^3 + x^4) \, dx}{\int_0^4 (4x - x^2) \, dx}$$

$$= \frac{\frac{1}{2}\left[\frac{16}{3}x^3 - 2x^4 + \frac{1}{5}x^5 + C\right]_0^4}{\left[2x^2 - \frac{1}{3}x^3 + C\right]_0^4} = 1.6$$

Thus the centroid has the coordinates (2, 1.6).

Example

Determine the position of the centroid of a semicircular sheet of uniform density and constant thickness.

The equation of a circle of radius r is

$$x^2 + y^2 = r^2$$

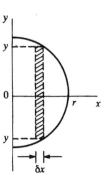

Fig. 8.12 Example

Consider the semicircle to be in the position shown in figure 8.12. Since the sheet is uniform and the area equally distributed about the x-axis then the centroid will lie along that axis. Consider the element of area indicated in the figure. It has an area $2y \, \delta x$. The total area of the semicircle is thus, as $\delta x \to 0$,

$$A = \int_0^r 2y \, dx$$

The moment of the element of area about the y-axis is $(2y\ \delta x)x$. Hence, as $\delta x \to 0$, the total moment is

$$\text{moment} = \int_0^r 2xy\ dx$$

Thus the centroid is at

$$\bar{x} = \frac{\int_0^r 2xy\ dx}{\int_0^r 2y\ dx}$$

But

$$y = \sqrt{r^2 - x^2}$$

Therefore

$$\bar{x} = \frac{\int_0^r 2x\sqrt{r^2 - x^2}\ dx}{\int_0^r 2\sqrt{r^2 - x^2}\ dx}$$

These two integrals can be integrated by the use of the substitution $u = \sqrt{(r^2 - x^2)}$ for the moment integral and $x = r \sin \theta$ for the area integral.

$$\bar{x} = \frac{\left[-\frac{2}{3}(r^2 - x^2)^{3/2} + C\right]_0^r}{\left[r^2(\theta - \frac{1}{2}\sin 2\theta) + C\right]_0^{\pi/2}} = \frac{\frac{2}{3}r^3}{\frac{1}{2}\pi r^2} = \frac{4r}{3\pi}$$

Thus the coordinates of the centroid are $(4r/3\pi, 0)$.

Example

Determine the position of the centroid of a sheet of uniform density and constant thickness and which has an area bounded by the graphs of the functions $y^2 = 8x$ and $y = x^2$.

Figure 8.13 shows the graphs of the two functions. The two curves intersect at $x^4 = 8x$, i.e. when $x = 0$ and 2 and consequently $y = 0$ and 4. Consider a strip of width δx. The area of the strip will be $[f(x) - g(x)]\ \delta x$. The total area bounded by the two graphs is thus

$$A = \int_0^2(\sqrt{8x} - x^2)\ dx = \left[\frac{2}{3}\sqrt{8}\,x^{3/2} - \frac{1}{3}x^3 + C\right]_0^2 = \frac{8}{3}$$

The moment of the strip about the y-axis is $[f(x) - g(x)]\ \delta x \times x$. Thus the total moment, when $\delta x \to 0$, is

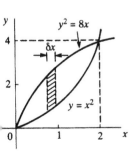

Fig. 8.13 Example

$$\text{moment} = \int_0^2 (\sqrt{8x} - x^2)x\, dx = \left[\tfrac{2}{5}\sqrt{8}\, x^{5/2} - \tfrac{1}{4}x^4 + C \right]_0^2 = 2.4$$

Hence

$$\bar{x} = \frac{2.4}{8/3} = 0.9$$

Now consider the distance of the centroid from the x-axis. The strip has an area $[f(x) - g(x)]\, \delta x$ effectively at its midpoint. This is a distance of $\tfrac{1}{2}[f(x) - g(x)]$ from the $y = x^2$ line and so a distance of $\tfrac{1}{2}[f(x) - g(x)] + x^2$ from the x-axis. Thus the moment of the strip about that axis is

$$\text{moment} = \left[\tfrac{1}{2}(\sqrt{8x} - x^2) + x^2 \right]\left(\sqrt{8x} - x^2 \right)\delta x$$

The moment about the x-axis for the entire area is thus, when we have $\delta x \to 0$,

$$\text{moment} = \int_0^2 \left[\tfrac{1}{2}(\sqrt{8x} - x^2) + x^2 \right]\left(\sqrt{8x} - x^2 \right) dx$$

$$= \int_0^2 \left(\sqrt{2x} + \tfrac{1}{2}x^2 \right)\left(2\sqrt{2x} - x^2 \right) dx$$

$$= \int_0^2 \left(4x - \tfrac{1}{2}x^4 \right) dx$$

$$= \left[2x^2 - \tfrac{1}{10}x^5 + C \right]_0^2 = \tfrac{24}{5}$$

Thus

$$\bar{y} = \frac{24/5}{8/3} = 1.8$$

Thus the centroid is at (0.9, 1.8).

Review problems

6 Determine the position of the centroid of a sheet which has an area bounded by the graph of the function $y = 4x$, the ordinates $x = 0$ and $x = 2$, and the x-axis.

7 Determine the position of the centroid of a sheet which has an area bounded by the graph of the function $y = x^2$, the ordinates $x = 1$ and $x = 3$, and the x-axis.

8 Determine the position of the centroid of a sheet which has an area bounded by the graph of the function $y = \sin x$, between 0 and π, and the x-axis.

Table 8.1 Centroids

Plane area		Centroid location
1	Rectangle	$\bar{x} = \frac{1}{2}b, \quad \bar{y} = \frac{1}{2}h$
2	Triangle	$\bar{x} = \frac{1}{3}(b+c), \quad \bar{y} = \frac{1}{3}h$
3	Right-angled triangle	$\bar{x} = \frac{1}{3}b, \quad \bar{y} = \frac{1}{3}h$
4	Semicircle	$\bar{y} = \dfrac{4r}{3\pi}$
5	Quarter circle	$\bar{x} = \bar{y} = \dfrac{4r}{3\pi}$
6	Parabolic semisegment	$\bar{x} = \frac{3}{8}b, \quad \bar{y} = \frac{2}{5}h$
7	Parabolic spandrel	$\bar{x} = \frac{3}{4}b, \quad \bar{y} = \frac{3}{10}h$

For the figures:

1. Rectangle: axes x, y; width b, height h.
2. Triangle: axes x, y; base b, height h, offset c.
3. Right-angled triangle: axes x, y; base b, height h.
4. Semicircle: axes x, y; radius r.
5. Quarter circle: axes x, y; radius r.
6. Parabolic semisegment: axes x, y; base b, height h; curve $y = h\left(1 - \frac{x^2}{b}\right)$.
7. Parabolic spandrel: axes x, y; base b, height h; curve $y = \frac{hx^2}{b}$.

9 Determine the position of the centroid of a sheet which has an area bounded by the graph of the function $y = x^3$, the ordinates $y = 0$ and $y = 1$, and the y-axis.

10 Determine the position of the centroid of a sheet in the form of the quadrant of a circle.

11 Determine the position of the centroid of the area bounded by the functions $y^2 = 4x$ and $x^2 = 4y$.

12 Determine the position of the centroid of the area bounded by the functions $y = x^2$ and $y = x$.

8.3.1 Centroids of composite areas

Many areas can often be considered as composite and made up of several parts, each part having a shape, such as a rectangle or circle, for which the positions of the centroids are known. The centroids of some commonly encountered plane areas are given in table 8.1.

For example, many beams have cross-sections which can be considered as composed of a number of areas. Thus for the L-shaped section shown in figure 8.14(a) we have effectively two rectangular areas A_1 and A_2. For rectangular areas the centroids are at the centres, where the diagonals cross. The first moment of area of area A_1 about the x-axis is $A_1 y_1$ and that for area A_2 is $A_2 y_2$. Thus the total moment of area about the x-axis is

$$\text{total moment} = A_1 y_1 + A_2 y_2$$

The total area is $A_1 + A_2$ and so the distance of the centroid of the composite section from the x-axis is

$$\bar{y} = \frac{A_1 y_1 + A_2 y_2}{A_1 + A_2} \qquad [16]$$

Similarly, considering the moments of the areas about the y-axis gives

$$\bar{x} = \frac{A_1 x_1 + A_2 x_2}{A_1 + A_2} \qquad [17]$$

It is possible to treat the absence of an area as a negative area. Thus the L-shaped section could have been considered as a large rectangle minus a rectangle in one corner, as illustrated in figure 8.14(b). The large rectangle has an area of A_3 with its centroid at its centre. It thus has a moment of area, about the x-axis, of $A_3 y_3$. The small, absent, area $-A_4$ has its centroid at its centre and thus a moment of area, about the x-axis, of $-A_4 y_4$. The total moment is thus

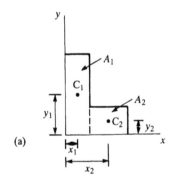

(a)

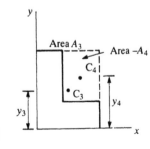

(b)

Fig. 8.14 L-shaped section

$$\text{total moment} = A_3y_3 - A_4y_4$$

The total area is $A_3 - A_4$. Hence the distance of the centroid of the L-shaped area from the x-axis is

$$\bar{y} = \frac{A_3y_3 - A_4y_4}{A_3 - A_4} \tag{18}$$

In a similar way the distance of the centroid from the y-axis can be determined.

Example

Determine the position of the centroid for the composite figure shown in figure 8.15.

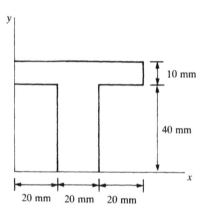

Fig. **8.15** Example

The shape is symmetric about the vertical axis through its centre and thus $\bar{x} = 30$ mm. We can consider the shape as either two rectangles, the top of the T and the stem, which are added or one large rectangle from which two smaller rectangles are subtracted. Considering it first as two rectangles added together, then for the moments of area about the x-axis we have a total moment of

$$60 \times 10 \times 45 + 40 \times 20 \times 20 = 43\,000 \text{ mm}^3$$

Hence, since the total area is $60 \times 10 + 40 \times 20 = 1400$ mm², the centroid is at a distance from the y-axis of

$$\bar{y} = \frac{43\,000}{1400} = 30.7 \text{ mm}$$

Thus the centroid is at (30 mm, 30.7 mm).

Now considering the shape as a large rectangle minus two smaller rectangles. The total moment of area is

$$50 \times 60 \times 25 - 20 \times 40 \times 20 - 40 \times 20 \times 20 = 43\ 000\ \text{mm}^3$$

i.e. the same total moment as derived by the first method.

Review problems

13 Determine the positions of the centroids for the composite sections shown in figures 8.16(a) and (b).

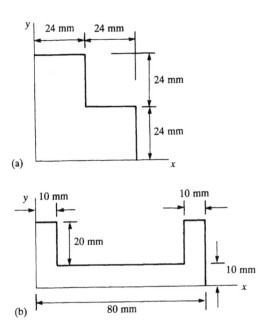

Fig. 8.16 Problem 13

8.3.2 The theorems of Pappus

There are two theorems of Pappus. The *first theorem* states that if a curve rotates round an axis, which does not cut the curve, then the area of the surface generated is equal to the length of the curve multiplied by the distance travelled by the centroid of the curve. This can be derived from a consideration of the curve in figure 8.17. The area swept out by an element of the curve δs when it makes one complete revolution about the x-axis is the surface area of a disc of radius y and so is $2\pi y\ \delta s$. The total area generated by the rotation is, when $\delta s \to 0$,

$$\text{area generated} = \int_0^s 2\pi y\ ds = 2\pi \int_0^s y\ ds$$

But the centroid of a line is given by equation [14] as

$$\bar{y} = \frac{\int_0^s y\ ds}{s}$$

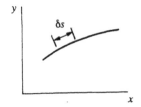

Fig. 8.17 A curve

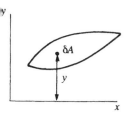

Fig. 8.18 An area

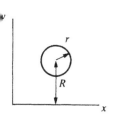

Fig. 8.19 Rotating a curve about the *x*-axis

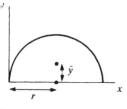

Fig. 8.20 Example

Hence

$$\text{area generated} = s \times 2\pi\bar{y} \qquad [19]$$

The *second theorem* states that if an area rotates round an axis, which does not cut the area, then the volume generated is equal to the area that rotates multiplied by the distance travelled by the centroid of the area. This can be derived from a consideration of figure 8.18. The volume generated by one revolution of the area δA about the *x*-axis is the volume of a ring of radius y, namely $2\pi y\,\delta A$. The total volume swept out by the rotation of the entire area is thus, when $\delta A \to 0$,

$$\text{volume generated} = \int_0^A 2\pi y\,dA = 2\pi \int_0^A y\,dA$$

But the centroid of an area is given by

$$\bar{y} = \frac{\int_0^A y\,dA}{A}$$

Hence

$$\text{volume generated} = A \times 2\pi\bar{y} \qquad [20]$$

Consider the first theorem in relation to the rotation of a circle of radius r about an axis which is a distance R from its centre (figure 8.19). The resulting shape is an anchor ring. The length of the rotating curve is the circumference of the circle, i.e. $2\pi r$. When rotated the centroid travels a distance $2\pi R$. Thus, applying the first theorem, the surface area of the anchor ring is

$$\text{area} = 2\pi r \times 2\pi R = 4\pi^2 rR$$

Now consider applying the second theorem, a circular area of πr^2 being rotated about an axis which is a distance R from its centre (figure 8.19 again). The resulting shape is an anchor ring. The distance travelled by the centroid is $2\pi R$. Thus, applying the second theorem, the volume of the anchor ring is

$$\text{volume} = \pi r^2 \times 2\pi R = 2\pi^2 r^2 R$$

Example

Determine the position of the centroid of a wire bent into a semicircle.

Figure 8.20 shows the wire. Because it is symmetric, the centroid must lie a distance r from the y-axis. Consider it being rotated about the x-axis. The area so generated will be the surface area of a sphere, namely $4\pi r^2$. The length of the wire is $2\pi r$. If the centroid is a distance $\bar{y}$ from that axis then it will rotate through $2\pi\bar{y}$. Thus, according to the first theorem,

$$4\pi r^2 = 2\pi r \times 2\pi\bar{y}$$

Hence $\bar{y} = 2r/\pi$.

Example

Determine the volume of the truncated cone shape shown in figure 8.21(a).

This figure can be considered as being generated by the rotation of the area shown in figure 8.21(b) about the y-axis. The area can be considered as a composite of a rectangle plus a triangle, hence

$$\text{area} = hr + \tfrac{1}{2}h(R - r) = \tfrac{1}{2}h(R + r)$$

The centroid of the area can be obtained by considering the composite. The moment of area of the rectangle about the y-axis is $hr \times \tfrac{1}{2}r$. The moment of area of the triangle about the y-axis is $\tfrac{1}{2}h(R - r) \times [\tfrac{1}{3}(R - r) + r]$, the centroid of a triangle being at one-third of its height. Thus the centroid is a distance from the y-axis of

$$\bar{y} = \frac{\tfrac{1}{2}h(R - r)[\tfrac{1}{3}(R - r) + r]}{\tfrac{1}{2}(R + r)}$$

Rotation of the centroid round the y-axis gives a distance travelled of $2\pi\bar{y}$. Hence the second theorem gives

$$\text{volume} = 2\pi\frac{\tfrac{1}{2}h(R - r)[\tfrac{1}{3}(R - r) + r]}{\tfrac{1}{2}(R + r)} \times \tfrac{1}{2}(R + r)$$

$$= \pi h(R - r)[\tfrac{1}{3}(R - r) + r]$$

Example

A cylinder, of diameter 200 mm, has a groove cut round its circumference. The groove is in the shape of a semicircle of diameter 30 mm. Determine the volume of material that has to be removed when the groove is cut.

Figure 8.22 shows the groove on the cylinder. We thus need to

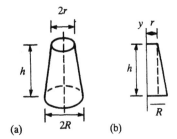

Fig. 8.21 Example

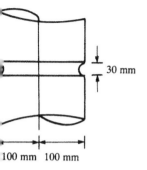

30 mm

100 mm   100 mm

Fig. 8.22 Example

determine the volume generated by the rotation of the semicircle about the y-axis. The semicircle has a centroid which is a distance of $4r/3\pi = 4 \times 15/3\pi = 6.37$ mm from its base. Thus the centroid is $100 - 6.37 = 93.63$ mm from the axis. The semicircle has an area of $\frac{1}{2}\pi r^2 = \frac{1}{2}\pi 15^2 = 353.43$ mm^2. Using the second theorem

$$\text{volume generated} = 353.43 \times 93.63 = 33\,092 \text{ mm}^3$$

Review problems

14 Determine the surface area of the truncated cone shape shown in figure 8.21.
   Hint: consider it to be generated by the rotation of a line, the centroid of the line being at its midpoint.
15 Determine the surface and volume generated when an equilateral triangle with sides of length L is rotated round its base.
16 Determine the volume of the anchor ring shape produced by rotating the circular region bounded by $(x - 2)^2 + y^2 = 1$ about the y-axis.
   Note: this is a circle of radius 1 with its centre at the co-ordinates (2, 0).
17 Determine the volume of material that will have to be removed from the rim of a disc, radius 40 mm, if a semicircular groove with a radius of 7.5 mm is to be machined from it.

8.4 Moments of inertia

Consider the rotation of a rigid body with a constant angular acceleration α about some axis (figure 8.23). Such a body can be considered to be made up of small elements of mass δm. For such an element at a distance r from the axis we will have a linear acceleration a at right angles to the direction of r. The force acting on the element and which is causing this acceleration must be given by $F = \delta m \times a$. But $a = r\alpha$. Thus

$$F = \delta m \times r\alpha$$

The moment of this force, the torque, about the axis is

$$\text{moment} = Fr = r^2\alpha\,\delta m$$

Thus the total torque T due to all the elements of mass in the body is

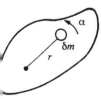

Fig. 8.23 Rotation of a rigid body

$$T = \sum r^2\alpha\,\delta m$$

In the limit as $\delta m \to 0$ we have

$$T = \int r^2 \alpha \, dm$$

Since α is a constant we can write

$$T = I\alpha$$

where I is termed the *moment of inertia* and is given by

$$I = \int r^2 \, dm \tag{21}$$

We can compare the equation $F = ma$ for the linear motion of a body with $T = I\alpha$ for the rotation of a body. The moment of inertia can be considered to play a role in angular motion comparable to the role of mass in linear motion. The moment of inertia is sometimes referred to as the *second moment of mass*. The term 'second' occurs because we have $r \times r$, while for what might be termed the first moment we would only have r.

It is often useful to discuss the moment of inertia of a body in terms of an equivalent, imaginary, body. If M is the total mass of a body then if we imagined all this mass to be concentrated at just one point a distance k from an axis we would have a moment of inertia of

$$I = Mk^2 \tag{22}$$

where k is called the *radius of gyration*.

Example

Determine the moment of inertia and radius of gyration of a uniform slender rod of length L about an axis through its centre and at right angles to its length.

Figure 8.24 shows the rod. Let m be the mass per unit length of the rod. Consider an element of mass δm a distance x from the axis. This element will have a moment of inertia about the axis of $(m \, \delta x) x^2$. Thus the total moment of inertia of the rod about the axis is

$$I = \sum_{x=-L/2}^{x=L/2} mx^2 \, \delta x$$

As $\delta x \to 0$ then

$$I = \int_{-L/2}^{L/2} mx^2 \, dx = m\left[\tfrac{1}{3}x^3 + C\right]_{-L/2}^{L/2} = \tfrac{1}{12}mL^3$$

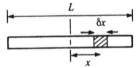

Fig. 8.24 Slender uniform rod

Since the total mass M of the rod is mL, then

$$I = \tfrac{1}{12}ML^2$$

If all the mass had been concentrated at one point a distance k from the axis then we would have had $I = Mk^2$. Thus the radius of gyration is

$$k = \sqrt{\tfrac{1}{12}L^2} = \sqrt{\tfrac{1}{12}}L$$

Example

Determine the moment of inertia and the radius of gyration of a uniform disc about an axis through its centre and at right angles to its plane.

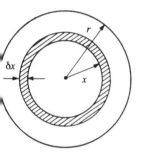

Fig. 8.25 Uniform disc

Figure 8.25 shows the disc. Let m be the mass per unit area. Consider a circular element of the disc with a radius x and width δx. The mass of the element will be $2\pi x\ \delta x \times m$. It will have a moment of inertia about the central axis of $(2\pi x\ \delta x \times m)x^2$. The total moment of inertia of the disc about the central axis is, when $\delta x \to 0$,

$$I = \int_0^r 2\pi mx^3\ dx = 2\pi m\left[\tfrac{1}{4}x^4 + C\right]_0^r = \tfrac{1}{2}\pi mr^4$$

Since the total mass M of the ring is $\pi r^2 m$, then

$$I = \tfrac{1}{2}Mr^2$$

If all the mass had been concentrated at one point a distance k from the axis then we would have had $I = Mk^2$. Thus the radius of gyration is

$$k = \sqrt{\tfrac{1}{2}}r$$

Example

Determine the moment of inertia and radius of gyration of a sphere about a diameter.

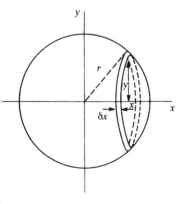

Fig. 8.26 Uniform sphere

Consider a sphere of radius r and mass per unit volume m and a thin slice of the sphere which is perpendicular to the diameter about which the moment of inertia is being determined (figure 8.26). The slice has a thickness of dx and is at a distance of x from the centre of the sphere. If the radius of the slice is y then it has a volume of $\pi y^2\ dx$. The mass of such a slice is thus $m\pi y^2\ dx$. The moment of inertia of a disc is given by $\tfrac{1}{2} \times$ mass $\times$ radius^2 (see the previous example). Hence the moment of inertia of the slice is $(m\pi y^2 \, dx)\frac{1}{2}y^2$. The moment of inertia of the sphere, when $\delta x \to 0$, is thus

$$I = \int_{-r}^{r} \tfrac{1}{2}m\pi y^4 \, dx$$

But $y^2 = r^2 - x^2$ and so

$$I = \int_{-r}^{r} \tfrac{1}{2}m\pi(r^2 - x^2)^2 \, dx = \tfrac{1}{2}m\pi \int_{-r}^{r} (r^4 - 2r^2x^2 + x^4) \, dx$$

$$= \tfrac{1}{2}m\pi \left[r^4 x - \tfrac{2}{3}r^2 x^3 + \tfrac{1}{5}x^5 \right]_{-r}^{r} = \tfrac{8}{15}m\pi r^5$$

But the mass of the sphere $M = \tfrac{4}{3}m\pi r^3$ and so

$$I = \tfrac{2}{5}Mr^2$$

If all the mass had been concentrated at one point a distance k from the axis then we would have had $I = Mk^2$. Thus the radius of gyration is

$$k = \sqrt{\tfrac{2}{5}} \, r$$

Review problems

18 Determine the moment of inertia and the radius of gyration of a uniform rectangular sheet about an axis through its centre and parallel to one side. Take the length to be $2L$ and the breadth to be $2b$, the moment of inertia being required parallel to the breadth.

19 Determine the moment of inertia and the radius of gyration of a uniform square sheet, side $2L$, about one side.

20 Determine the moment of inertia and the radius of gyration of a uniform right circular cone about its axis.
    Hint: treat the cone as a solid generated by revolution about the x-axis, then $y/x = r/h$, where r is the cone radius and h its height.

8.4.1 Perpendicular and parallel axes theorems

There are often situations where the moment of inertia is known for a body about a particular axis and the moment of inertia is then required about some other axis. This other axis may be at right angles or perhaps parallel to the axis for which the moment of inertia is known. There are two theorems, the perpendicular axes theorem and the parallel axes theorems, which enable such conversions to be easily made.

Consider an object with a plane area and an element in that

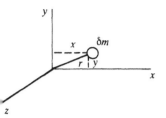

Fig. 8.27 Perpendicular axes

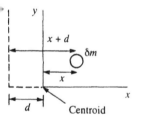

Fig. 8.28 Parallel axes

area with a mass δm (figure 8.27). The moment of inertia of the element about an axis at right angles to the plane, the z-axis, and a distance r from the element, is $r^2\,\delta m$. The moment of inertia of the plane area about the z-axis is thus

$$I_z = \int r^2\,dm$$

But the coordinates of the element with respect to the x and y-axes are x and y, where $r^2 = x^2 + y^2$. Thus

$$I_z = \int (x^2 + y^2)\,dm = \int x^2\,dm + \int y^2\,dm$$

But the moment of inertia about the x-axis is

$$I_x = \int x^2\,dm$$

and the moment of inertia about the y-axis is

$$I_y = \int y^2\,dm$$

Hence

$$I_z = I_x + I_y \qquad\qquad [23]$$

This is referred to as the *perpendicular axes theorem*.

Consider an element of mass δm in a plane area object, with the element being a distance x from the y-axis through the centroid of the area (figure 8.28). The moment of inertia of that element about such an axis is $x^2\,\delta m$ and so the moment of inertia of the entire object about the y-axis through the centroid is

$$I_c = \int x^2\,dm$$

Now consider another y-axis which is parallel to the one through the centroid and displaced from it by a distance d. The element δm is now a distance of $(x + d)$ from this axis and so the moment of inertia of the element about the axis is $(x + d)^2\,\delta m$. The moment of inertia of the entire object about this new y-axis is thus

$$I_d = \int (x + d)^2\,dm$$

$$= \int x^2\,dm + 2d\int x\,dm + d^2\int dm$$

The first integral is the moment of inertia about the centroid I_c. The second integral involves the product of an element of mass and its distance from the centroid. It is thus the sum of the moments of all the elements of mass about the centroid. But this sum must be zero. Hence the second integral is zero. The third integral involves the sum of all the elements of mass and so is M, the total mass of the object. Hence

$$I_d = I_c + Md^2 \qquad [24]$$

This is the *parallel axes theorem*. The parallel axes theorem is particularly useful for determining the moments of inertia of composite objects (see the next section in this chapter).

Example

The moment of inertia of a uniform rectangular object about an axis parallel to one side and through its centroid is $ML^2/3$. What is the moment of inertia of the rectangle about the base of the rectangle?

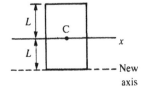

Fig. 8.29 Example

Figure 8.29 shows the rectangle. The centroid is at the centre of it and the axis about which the moment of inertia is $ML^2/3$ is the x-axis. The moment of inertia is required about the base. This is an axis which is parallel to the x-axis and displaced from it by a distance L. Hence, using the parallel axes theorem, the required moment of inertia is

$$\tfrac{1}{3}ML^2 + ML^2 = \tfrac{4}{3}ML^2$$

Example

The moment of inertia of a uniform disc is $Mr^2/4$ about a diameter. Determine the moment of inertia about an axis at right angles to the plane of the disc and through its centroid, the z-axis in figure 8.30.

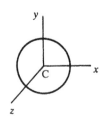

Fig. 8.30 Example

The disc is in the xy plane. The moment of inertia of the disc about the x-axis is $Mr^2/4$. Likewise, the moment of inertia of the disc about the y-axis is $Mr^2/4$. Using the perpendicular axes theorem,

$$I_z = \tfrac{1}{4}Mr^2 + \tfrac{1}{4}Mr^2 = \tfrac{1}{2}Mr^2$$

Review problems

21 Determine the moment of inertia of a uniform disc about a tangent to the disc, given that the moment of inertia about a diameter is $Mr^2/4$.
22 Determine the moment of inertia of a uniform lamina in the shape of an isosceles triangle of height h about (a) an axis through the centroid and parallel to the base, (b) the base. The moment of inertia of the triangle about an axis through the vertex and parallel to the base is $Mh^2/2$.

8.4.2 Composite bodies

Many objects can be considered as composite, having elements which are standard shapes for which the moments of inertia are known. Table 8.2 gives moments of inertia of common shapes. The moment of inertia of the composite body about some axis can then be obtained by adding algebraically the moments of inertia of each element about the axis. For this purpose the theorem of parallel axes is useful in converting the moments of inertia of standard shapes about axes through their centroids into moments of inertia about the axis required for the composite body.

Example

An assembly consists of identical solid spheres, mass M and radius r, at each end of a horizontal rod of length $2L$ and negligible mass. Determine the moment of inertia of the assembly about a vertical axis through the centre of the rod.

Each sphere has a moment of inertia of $\frac{2}{5}Mr^2$ about an axis through its centre. The centre of each sphere is a distance of $r + L$ from the central axis. Hence the moment of inertia of a sphere about the central axis is, using the theorem of parallel axes,

$$\frac{2}{5}Mr^2 + M(r+L)^2$$

Since there are two spheres, then the moment of inertia of the assembly about the central axis is

$$2\left[\frac{2}{5}Mr^2 + M(r+L)^2\right]$$

Example

Determine the moment of inertia of a circular plate about an axis perpendicular to the plate and passing through its centre if the plate has a radius R and a central hole of radius r. The plate has a uniform thickness t and a density ρ.

Consider the object as a composite which consists of a circular plate of radius R and another of radius r, this smaller diameter plate having a negative moment of inertia because it is a hole. The moment of inertia of a disc about an axis at right angles to its plane and through its centre is $\frac{1}{2}MR^2$. Thus the moment of inertia of the composite is

$$\frac{1}{2}MR^2 - \frac{1}{2}mr^2$$

Table 8.2 Moments of inertia

Uniform solid		Moment of inertia
1	Slender rod	$I_x = I_y = \frac{1}{12}ML^2, \quad I_z = 0$
2	Sphere	$I_x = I_y = I_z = \frac{2}{5}Mr^2$
3	Thin circular disc	$I_x = I_y = \frac{1}{4}Mr^2, \quad I_z = \frac{1}{2}Mr^2$
4	Cylinder	$I_x = I_y = \frac{1}{12}M(3r^2 + h^2), \quad I_z = \frac{1}{2}Mr^2$
5	Thin plate	$I_x = \frac{1}{12}Mb^2, \quad I_y = \frac{1}{12}Ma^2, \quad I_z = \frac{1}{12}M(a^2 + b^2)$
6	Cone	$I_x = I_y = \frac{3}{80}(4r^2 + h^2), \quad I_z = \frac{3}{10}Mr^2$

But $M = \pi R^2 t\rho$ and $m = \pi r^2 t\rho$. Thus the moment of inertia of the composite is

$$\tfrac{1}{2}\pi R^2 t\rho R^2 - \tfrac{1}{2}\pi r^2 t\rho r^2 = \tfrac{1}{2}\pi t\rho(R^4 - r^4)$$

Review problems

23 An object is in the form of a uniform solid hemisphere of radius *r* with its plane face joined to the base of a uniform circular cylinder of radius *r* and height 2*r*. Both elements have the same density ρ. Determine the moment of inertia of the composite body about an axis through its centre of gravity.
Note: you will need to determine the position of the centre of gravity.

24 An object consists of a uniform rod of length 250 mm and mass 0.2 kg. At each end of the rod are spheres. At one end the sphere has a radius of 60 mm and a mass of 0.4 kg and at the other end a radius of 80 mm and a mass of 0.6 kg. Determine the moment of inertia of the object about an axis at right angles to the rod and passing through its centre of gravity.
Note: you will need to determine the position of the centre of gravity.

8.5 Second moment of area

There are many other situations in engineering where integrals of the form

$$\int y^2 \, dA$$

are involved, e.g. in the bending of beams. Such integrals are similar to the mass moment of inertia integrals discussed in section 8.4. For this reason they are sometimes referred to as the *area moment of inertia*. A more usual term is, however, the *second moment of area*. The first moment of area involves the integral of *y* d*A*, the second moment *y* × *y* d*A*.

The second moments of area can be derived in a similar way to the mass moments of inertia obtained in section 8.4. The theorems of perpendicular and parallel axes (section 8.4.1) also apply. Thus, the second moment of area about the *z*-axis which is mutually at right angles from the *x*- and *y*-axes is given by

$$I_z = I_x + I_y \qquad\qquad [25]$$

The second moment of area about an axis which is parallel to, and a distance *d* from, the axis through the centroid is given by

$$I_d = I_c + Ad^2 \qquad\qquad [26]$$

Composite areas can be determined in the same way as the mass moment of inertia of composite solids referred to in section 8.4.2.

The *radius of gyration* of a planar area is often used. The radius of gyration *k* about an axis is the square root of the second moment of area about that axis divided by the square root of the plane area, i.e.

$$k = \sqrt{\frac{I}{A}} \qquad\qquad [27]$$

The equation is similar to that used for the radius of gyration with mass moments of inertia.

Table 8.3 gives second moments of area of a number of commonly encountered plane areas.

Example

Determine the second moment of area of a rectangular area about an axis through the centroid and parallel to one side.

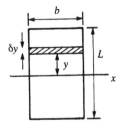

Fig. 8.31 Example

Figure 8.31 shows the area and the x-axis about which the second moment of area is required. Consider a strip of width δy a distance y from the x-axis. The strip has an area $\delta A = b\,\delta y$. The second moment of area about the axis is $y^2 b\,\delta y$. Hence the second moment of area of the entire rectangle about the axis is the sum of all these strips and hence, when $\delta y \to 0$,

$$I = \int_{-L/2}^{L/2} y^2 b\,dy = b\left[\tfrac{1}{3}y^3 + C\right]_{-L/2}^{L/2} = \tfrac{1}{12}bL^3$$

Example

Determine the second moment of area of the rectangular area considered in the previous example, but this time about an axis through the centroid but at right angles to the plane of the area.

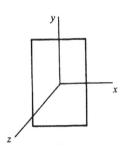

Fig. 8.32 Example

The situation is as shown in figure 8.32. The second moment of area is required about the z-axis. The second moment of area about the x-axis is, as obtained in the previous example, $bL^3/12$. The second moment of area about the y-axis will be similar to that about the x-axis and so is $Lb^3/12$. The second moment of area about the z-axis is given by the perpendicular axes theorem (equation [25] as

$$I_z = I_x + I_y = \tfrac{1}{12}bL^3 + \tfrac{1}{12}Lb^3$$

Example

Determine the second moment of area of the rectangular area considered in the two previous examples, but this time about the base of the rectangle.

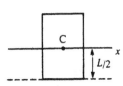

Fig. 8.33 Example

The situation is as shown in figure 8.33. The second moment of area about an axis through the centroid is $bL^3/12$. The second

Table 8.3 Second moments of area

	Plane area		Second moment of area
1	Rectangle		$I_x = \frac{1}{12}bh^3, \quad I_y = \frac{1}{12}hb^3$
2	Circle		$I_x = I_y = \frac{1}{4}\pi r^4$
3	Semicircle		$I_x = I_y = \frac{1}{8}\pi r^4$
4	Quarter circle		$I_x = I_y = \frac{1}{16}\pi r^4$
5	Triangle		$I_x = \frac{1}{36}bh^3, \quad I_y = \frac{1}{36}bh(b^2 - bc + c^2)$
6	Right-angled triangle		$I_x = \frac{1}{36}bh^3, \quad I_y = \frac{1}{36}hb^3$

moment of area is required about an axis which is parallel to this and a distance $L/2$ from it. Hence, using the parallel axes theorem (equation [26]) we have

$$I = \frac{1}{12}bL^3 + A\left(\frac{L}{2}\right)^2 = \frac{1}{12}bL^3 + bL \times \frac{1}{4}L^2 = \frac{1}{3}bL^3$$

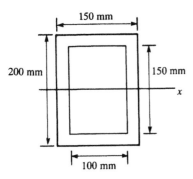

Fig. 8.34 Example

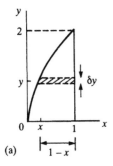

(a)

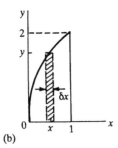

(b)

Fig. 8.35 Example

Example

Determine the second moment of area of the rectangular tube section shown in figure 8.34 about an axis through its centroid and parallel to its base.

We can consider the section as being an area of 200 by 150 mm minus 100 by 150 mm. The second moment of an area about an axis through its centroid and parallel to its base is $bL^3/12$. Thus the second moment of area is

$$I = \tfrac{1}{12}150 \times 200^3 - \tfrac{1}{12}100 \times 150^3 = 7.19 \times 10^7 \text{ mm}^4$$

Example

Determine the second moment of area of the area bounded by $y^2 = 4x$, the ordinate $x = 1$ and the x-axis about (a) the x-axis, (b) the y-axis.

(a) Figure 8.35(a) shows the area concerned. Consider a slice of width δy a distance y from the x-axis. If x is the point at which this slice intersects with the curve, then the length of the slice will be $(1 - x)$. The area of the slice is thus $\delta A = (1 - x)\delta y$. The second moment of area of the slice about the x-axis is then $y^2[(1 - x)\delta y]$. Hence the second moment of the bounded area about the x-axis is

$$I_x = \int_0^2 y^2(1-x)\,dy = \int_0^2 y^2(1-\tfrac{1}{4}y^2)\,dy = \left[\tfrac{1}{3}y^3 - \tfrac{1}{20}y^5 + C\right]_0^2$$

$$= 1.07 \text{ units}^4$$

(b) For the second moment of area about the y-axis, consider the slice shown in figure 8.35(b). The area of the slice $\delta A = y\,\delta x$. The slice will have a second moment of area about the y-axis of $x^2(y\,\delta x)$. The second moment of area for the bounded area is thus

$$I_y = \int_0^1 x^2 y\,dx = \int_0^1 x^2(2\sqrt{x})\,dx = \left[\tfrac{4}{7}x^{7/2} + C\right]_0^1$$

$$= 0.57 \text{ unit}^4$$

Review problems

25 Determine the second moment of area of:
   (a) a circle of radius r about a diameter,
   (b) a semicircle of radius r about an axis through its centroid and parallel to its diameter,
   (c) a circular tube section with an internal diameter of d and an external diameter D about an axis through a diameter.
   Hint: you might find it easier to determine the second moment

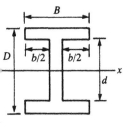

Fig. 8.36 Problem 26

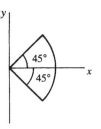

Fig. 8.37 Problem 28

Further problems

of area for (b) about the base of the semicircle and then use the theorem of parallel axes to obtain the moment through the centroid and for (c) as a composite.

26 Determine the second moment of area of the I-section shown in figure 8.36 about the axis through its centroid and in the plane of the section.

27 Determine the second moment of area of a circular area of radius 20 mm about an axis which is parallel to a diameter and a distance of 50 mm from the centre of the circle.

28 Determine the radius of gyration of the quarter circle area shown in figure 8.37 about the y-axis. The circle has a radius of 200 mm and is described by the equation $x^2 + y^2 = 200^2$.

29 A composite area consists of a semicircle of radius 70 mm attached at its base to a rectangular area of length 140 mm and width 30 mm. Determine the second moment of area of the composite area about the common base.

30 Determine the distance of the centre of gravity from one end of a uniform cross-section rod if the density of the rod is proportional to the distance from the end.

31 Determine the distance of the centre of gravity from the lighter end of a uniform cross-section rod if the mass per unit length of the rod a distance x from the lighter end is $(1 + x/2)$ kg/m and the rod is 6 m long.

32 Determine the centre of mass of the right circular cone which is generated by the rotation of the line $y = mx$, between $x = 0$ and $x = h$, about the x-axis.

33 Determine the position of the centre of gravity of a cylinder of diameter 40 mm and height 60 mm and in which a flat-ended central hole of diameter 30 mm and depth 50 mm has been cut.

34 Determine the position of the centroid of a sheet which has an area bounded by the graph of the function $y = 4 - x^2$ and the x-axis.

35 Determine the position of the centroid of a sheet which has an area bounded by the graph of the function $y = x^3$, the ordinates $y = 0$ and $y = 9$, and the y-axis.

36 Determine the position of the centroid of a sheet which has an area bounded by the graph of the function $y = 3x^2$, the ordinates $x = 0$ and $x = 3$, and the x-axis.

37 Determine the position of the centroid of a sheet which has an area bounded by the graph of the function $2y = x^2$, the ordinates $x = 0$ and $x = 2$, and the x-axis.

38 Determine the position of the centroid of the area bounded by the functions $y^2 = 4x$ and $y = 3x$.

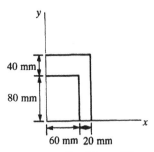

Fig. 8.38 Problem 40

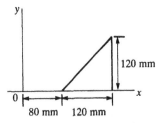

Fig. 8.39 Problem 43

39 Determine the position of the centroid of the area bounded by the functions $y^2 = x$ and $y = x$.

40 Determine the position of the centroid for the plate shown in figure 8.38 by considering it as a composite figure.

41 Using the theorems of Pappus, determine the surface area and the surface volume of the solid formed by the rotation of a square of side L about an axis in its plane through a corner which is at right angles to the diagonal through the corner.

42 Using the theorems of Pappus, determine the surface areas and volumes of a cylinder.

43 Using the Pappus theorem, determine the volume of the solid generated by rotating the right triangle shown in figure 8.39 by one complete revolution about the x-axis.

44 Determine the moment of inertia and the radius of gyration of an isosceles triangle about an axis through its vertex and at right angles to its base.

45 Determine the moment of inertia and the radius of gyration of a circular cylinder about its axis.

46 Determine the moment of inertia and the radius of gyration of a flat circular ring with an inner radius of r and an outer radius of $2r$, about an axis through its centre and at right angles to its plane.

47 Determine the moment of inertia of a uniform square sheet, side of length $2L$, about an axis perpendicular to the plane of the square and at one corner. The moment of inertia about an axis through the centroid and parallel to a side is $ML^2/3$.

48 Determine the moment of inertia of a uniform rod of length $2L$ about an axis perpendicular to one end of the rod.

49 An object consists of a solid hemisphere of radius r with a solid right circular cone of radius r attached so that its base is attached to the plane surface of the hemisphere. The cone is solid and has a height of r. Determine the moment of inertia of the composite about an axis in the plane of the common bases.

50 Determine the second moment of area of:
(a) a non-right-angled triangle of base b and height h about an axis through its centroid and parallel to its base,
(b) a right-angled triangle of base b and perpendicular height h about an axis through its centroid and parallel to its base,
(c) a right-angled triangle of base b and perpendicular height h about an axis through its centroid and at right angles to its base.

51 Determine the second moment of area of the T-section shown in figure 8.40 about an axis parallel to the top of the T and through its centroid.

52 Determine the second moment of area of a rectangular area, in which a circular hole has been drilled, about its base. The rectangle has a width of 100 mm and a height of 150 mm. The hole has a diameter of 50 mm and is centrally situated in the rectangular area.

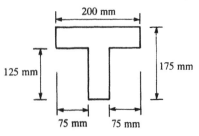

Fig. 8.40 Problem 51

53 Determine the second moment of area of a semicircular area, which has had a semicircular area removed from it, about its diameter. The semicircular area has a radius of 60 mm and the area removed has a radius of 20 mm and a centre and base coinciding with that the larger semicircle.

9 Mean values

If we have a set of values of some quantity then the average, or mean, value is the sum of the set of values divided by the number of quantities in the set. For example, if the resistance of four resistors is 10.0 Ω, 10.4 Ω, 10.2 Ω and 10.6 Ω, then the mean value of the resistance is

$$\text{mean resistance} = \frac{10.0 + 10.4 + 10.2 + 10.6}{4} = 10.3 \ \Omega$$

Now suppose we have a current i which varies continuously with time t, i.e. the current is a function of time $i = f(t)$, and we require the mean value of the current over some time interval. We can obtain the mean value in a similar way to the mean value of the discrete resistances. All we need to do is take the values of the current at a number of instances during the time interval and divide the sum of the current values by the number of values considered. In general, we would expect to improve the accuracy of the mean by taking a large number of current values since then we would be less likely to miss any current fluctuations and not take them into account.

This chapter is about how we can obtain the mean value of functions. This might, for example, be the mean velocity of an object where the velocity is some function of time. Such means are considered in section 9.1.1. In the case of alternating currents the concern is often the root-mean-square value, i.e. the square root of the mean values of the square of the current function. Root-mean-square values are considered in section 9.2.

9.1.1 The mean value of a function

Consider the problem of determining the mean value of a function $y = f(x)$ for values of x from a to b, as illustrated in figure 9.1. The mean value will be the mean value of the ordinates of the graph

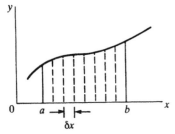

Fig. 9.1 $y = f(x)$

between $x = a$ and $x = b$. Suppose, therefore, we divide the area under the graph between these ordinates into a number of equal width strips. If we have n such strips, each of width δx, then if the values of the mid-ordinates of the strips are $y_1, y_2, y_3, \ldots y_n$ we have a mean value $\bar{y}$ of

$$\bar{y} = \frac{y_1 + y_2 + y_3 + \ldots + y_n}{n}$$

If we multiply the numerator and denominator by δx, then

$$\bar{y} = \frac{y_1 \delta x + y_2 \delta x + y_3 \delta x + \ldots + y_n \delta x}{n \delta x}$$

But $n \, \delta x$ is $b - a$. Hence, in the limit as $\delta x \to 0$, we have

$$\bar{y} = \frac{1}{b-a} \int_a^b y \, dx \qquad\qquad [1]$$

Since the sum of all the $y \, \delta x$ terms is the area under the graph between $x = a$ and $x = b$, i.e. the integral gives the area, then we can also write equation [1] as

$$\bar{y} = \frac{\text{area under graph}}{b-a}$$

We can rewrite this as

$$\bar{y} \times (b-a) = \text{area under the graph}$$

But $\bar{y} \times (b-a)$ is the area of a rectangle of height $\bar{y}$ and width $(b-a)$. Such a rectangle has an area equal to that under the curve between the limits of a and b, as illustrated in figure 9.2. The horizontal line at the mean value of y is a line through the curve such that the areas between it and the curve balance, i.e. there is as large an area between it and the curve above it as below it.

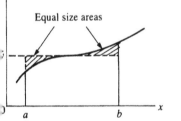

Equal size areas

Fig. 9.2 Area under the curve

Example

Determine the mean value of the function $y = \sin x$ between $x = 0$ and $x = \pi$.

Using equation [1],

$$\bar{y} = \frac{1}{b-a} \int_a^b y \, dx = \frac{1}{\pi - 0} \int_0^\pi \sin x \, dx = \frac{1}{\pi} [-\cos x + C]_0^\pi$$

$$= \frac{2}{\pi} = 0.637$$

Example

An object is shot vertically upwards with a velocity v metre/second which is a function of time t seconds. If $v = 15 - 10t$, determine the mean velocity over the time interval $t = 0$ to $t = 1$ s.

Using equation [1],

$$\bar{v} = \frac{1}{b-a} \int_a^b v \, dt = \frac{1}{1-0} \int_0^1 (15 - 10t) \, dt$$

$$= [15t - 5t^2 + C]_0^1 = 10 \text{ m/s}$$

Review problems

1 Determine the mean values of the following functions between the specified limits:
   (a) $y = x^2$ between $x = 1$ and $x = 4$,
   (b) $y = e^x$ between $x = 1$ and $x = 4$,
   (c) $y = 3x^2 + 1$ between $x = 1$ and $x = 2$,
   (d) $y = x + 1$ between $x = 0$ and $x = 2$,
   (e) $y = \sqrt{(x + 1)}$ between $x = 0$ and $x = 3$,
   (f) $y = 2 + 3 \sin t$ between $t = 0$ and $t = \pi$

2 In simple harmonic motion the displacement x of an object from its rest position is a function of time t and given by the equation $x = A \cos \omega t$. Determine the mean value of the displacement during one quarter of a complete oscillation, i.e. from $\omega t = 0$ to $\omega t = \pi/2$.

3 An alternating voltage v varies with time t according to the equation $v = 10 \sin \omega t$. Determine the mean value of the voltage over half a cycle, i.e. from $\omega t = 0$ to $\omega t = \pi$.

4 Determine the mean value of y for a semicircle drawn on the x-axis and having a radius r. Note: for a circle $x^2 + y^2 = r^2$.

5 The number of atoms N remaining in a radioactive material after a time t is given by the equation $N = N_0 e^{-\lambda t}$, where N_0 and λ are constants. Determine the mean number of radioactive atoms in the sample during the period $t = 0$ to $t = 1/\lambda$.

6 The current through a component in a circuit is given by $i = I \sin(\omega t - \phi)$ when the potential difference across it is $v = V \sin \omega t$. Determine the mean value of iv in the time $t = 0$ to $t = 2\pi/\omega$.

9.2 The root-mean-square value The power dissipated by an alternating current i when passing through a resistance R is $i^2 R$. Since i varies with time then the power will vary with time. The mean power dissipated during one cycle, i.e. a time equal to the period T, is thus

$$\text{mean power} = \frac{1}{T} \int_0^T i^2 R \, dt = \frac{R}{T} \int_0^T i^2 \, dt$$

In terms of power dissipation the effective value of an alternating current can be specified in terms of the value of the direct current which, when flowing through a resistor, dissipates the same power. The power dissipated by the direct current I_{eff} to give the same power as the mean power is $I_{eff}^2 R$. Thus we must have

$$I_{eff}^2 = \frac{1}{T}\int_0^T i^2 \, dt$$

Thus the effective current is

$$I_{eff} = \sqrt{\frac{1}{T}\int_0^T i^2 \, dt}$$ [2]

This is known as the *root-mean-square current*. In a similar way we can derive the equation for the *root-mean-square voltage*.

$$V_{eff} = \sqrt{\frac{1}{T}\int_0^T v^2 \, dt}$$ [3]

Thus to determine the root-mean-square value of a function we:

1 Square the function.
2 Determine the mean value of the squared function over the required interval.
3 Take the square root of this mean value.

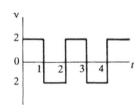

(a)

Example

Determine the root-mean-square value of the voltage giving the rectangular waveform shown in figure 9.3(a) over a period. The waveform has a period of $t = 2$ s and the voltages are in volts.

Squaring the function results in the graph shown in figure 9.3(b). Thus $v^2 = 4$. Hence the root-mean-square value is given by equation [3] as

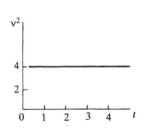

(b)

Fig. 9.3 Example

$$V_{rms} = \sqrt{\frac{1}{T}\int_0^T v^2 \, dt} = \sqrt{\frac{1}{2}\int_0^2 4 \, dt} = \sqrt{\frac{1}{2}[4t + C]_0^2} = 2 \text{ V}$$

Example

Determine the root-mean-square value of the voltage waveform shown in figure 9.4. The waveform has a period of $t = 2$ s and the voltages are in volts.

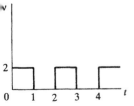

Fig. 9.4 Example

For the first half of the period we have $v^2 = 4$ V. For the second half of the period period we have $v^2 = 0$. The period is 2 s. Thus the root-mean-square value is given by

$$V_{rms} = \sqrt{\frac{1}{T}\int_0^T v^2 \, dt} = \sqrt{\frac{1}{2}\left(\int_0^1 4 \, dt + \int_1^2 0 \, dt\right)}$$

$$= \sqrt{\frac{1}{2}[4t + C]_0^1} = \sqrt{2} \text{ V}$$

Example

Determine the root-mean-square value of an alternating current described by the equation $i = I \sin \omega t$.

The period T is $2\pi/\omega$. Thus, using equation [2], the root-mean-square current is

$$I_{rms} = \sqrt{\frac{1}{T}\int_0^T i^2 \, dt}$$

$$= \sqrt{\frac{1}{T}\int_0^T I^2 \sin^2 \omega t \, dt}$$

$$= I\sqrt{\frac{\omega}{2\pi}\int_0^{2\pi/\omega} \frac{1}{2}(1 - \cos 2\omega t) \, dt}$$

$$= I\sqrt{\frac{\omega}{4\pi}\left[t - \frac{1}{2\omega}\sin 2\omega t + C\right]_0^{2\pi/\omega}}$$

$$= \frac{I}{\sqrt{2}}$$

Review problems

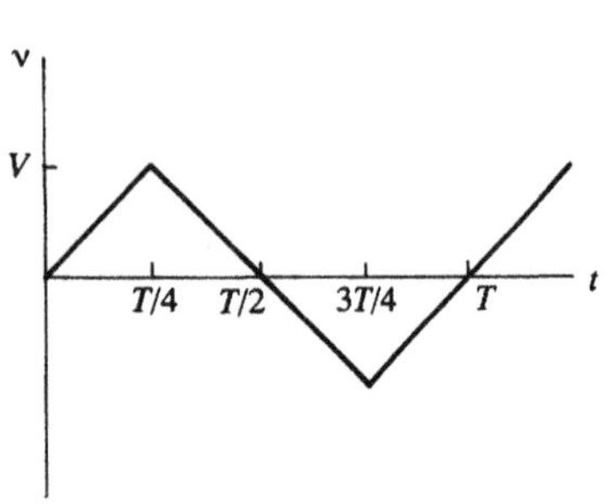

7 Determine the root-mean-square value of the function $y = t^2$ over the interval $t = 1$ to $t = 3$.

8 Determine the root-mean-square value of the current described by $i = 1 + \sin t$ over the time interval $t = 0$ to $t = 2\pi$.

9 Determine the root-mean-square value of the voltage $v = \sin 2t$ over the time interval $t = 0$ to $t = \pi$.

10 Determine the root-mean-square value of the half-rectified voltage sine wave shown in figure 9.5. Between the times $t = 0$ and $t = \pi/\omega$ the equation is $v = V \sin \omega t$, between $t = \pi/\omega$ and $2\pi/\omega$ we have $v = 0$.

11 Determine the root-mean-square value of the periodic triangular voltage shown in figure 9.6.
Hint: because the result of squaring each quarter period is the same you only need to determine the mean values of the squares for a quarter period.

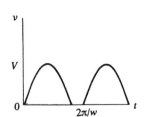

Fig. 9.5 Problem 10

Fig. 9.6 Problem 11

Further problems

12 Determine the mean values of the following functions between the specified limits:
(a) $y = 6x^2$ between $x = 1$ and $x = 3$,
(b) $y = 4x + 1$ between $x = 0$ and $x = 2$,
(c) $y = 10 \sin 2x$ between $x = 0$ and $x = \pi$,
(d) $y = 1/x$ between $x = 1$ and $x = 3$,
(e) $y = \sin \omega t + \cos \omega t$ between $t = 0$ and $t = 1$,
(f) $y = x \sin x$ between $x = 0$ and $x = \pi$

13 An alternating voltage v varies with time and is described by the equation $v = 10 \sin 250t$. Determine the mean value of the voltage between $t = 0$ and $t = \pi/100$.

14 When an object is fired as a projectile with an initial velocity u at an angle θ to the horizontal then the horizontal range x is given by

$$x = \frac{u^2 \sin 2\theta}{g}$$

If the angle at which the projectile is fired can vary between 0 and $\pi/2$, what is the mean value of the range?

15 The velocity v of an object performing simple harmonic motion varies with time t and is described by $v = 10 \sin \omega t$. Determine the mean velocity between the times $t = 0$ and $t = \pi/\omega$.

16 Determine the mean value of y for the graph of the parabola $y^2 = 4ax$ between $x = 0$ and $x = 4a$.

17 Determine the root-mean-square value of the function of time $y = A \sin(\omega t + \phi)$ over the time interval $t = 0$ to $t = 2\pi/\omega$.

18 Determine the root-mean-square value of the current $i = 2 + 3t$ over the time interval $t = 0$ to $t = 2$.

19 Determine the root-mean-square value of the current

$$i = I_1 \sin(\omega t + \phi_1) + I_2 \sin(2\omega t + \phi_2)$$

over a period $T = 2\pi/\omega$.

20 Determine the root-mean-square value for a full wave rectified sinusoidal voltage.
Hint: since the waveform is the same in each half period, find the mean of the square of the voltage over half a period, i.e. 0 to π/ω. The waveform in half a period can be assumed to be $v = V \sin \omega t$.

21 Determine the root-mean-square value for the periodic waveform shown in figure 9.7. The period is 4 s.

22 Determine the root-mean-square values for the following:
(a) $y = 3x$ from $x = 0$ to $x = 4$,
(b) $y = 5 \sin 2x$ from $x = 0$ to $x = \pi$,
(c) $y = 2 \sin x$ from $x = 0$ to $x = 2\pi$,
(d) $y = 2x - x^2$ from $x = 0$ to $x = 2$,
(e) $y = 4 + 3 \sin x$ from $x = 0$ to $x = 2\pi$

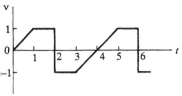

Fig. 9.7 Problem 21

10 Numerical integration

10.1 Numerical integration

In engineering there are often situations where the functions that have to be integrated are only specified by a set of values. This might be, for example, in the computation of the work done as the integral of $F\,dx$ when a force F moves its point of application through a distance x, the force–distance values having been obtained experimentally. In such cases numerical integration can be used. Also there are some functions which defy the usual methods of integration, there being no elementary function which has the function as their derivative. For example, there are no functions which have the following as their derivatives:

$$\frac{1}{\ln x}, \quad \sin x^2, \quad \frac{\cos x}{x}, \quad \sqrt{1 - x^3}$$

Numerical integration methods can be used, however complex a function is.

Computer programs for the integration of functions whose form is known use numerical techniques of integration rather than analytical methods. These enable a general technique to be used which can be applied to all integrals, whatever their form or however complex they might be.

The simplest way of determining the area under a graph between ordinates, i.e. obtaining the definite integral for the function, is to plot the function on squared paper and count the squares, units of area, under the graph. This can be applied without any knowledge of the function concerned. More sophisticated techniques are the mid-ordinate rule, the trapezium rule and Simpson's rule. These techniques are based on replacing the function between two values of the variable by a polynomial series which can be easily integrated numerically, the essential difference between the techniques being the number of terms in the polynomial expression. Thus, in general, we can write

$$f(x) = A + Bx + Cx^2 + Dx^3 + \ldots$$

At its simplest the function being integrated is replaced between two ordinates by a straight line with the equation $y = A$, i.e. just a constant. A more complex replacement is a straight line with the equation $y = A + Bx$ with a yet more complex being the curve with the equation $f(x) = A + Bx + Cx^2$. This chapter is a consideration of such techniques.

10.2 Mid-ordinate rule

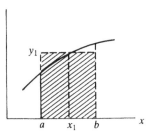

Fig. 10.1 Mid-ordinate approximation

Consider the function $y = f(x)$ which is described by the graph shown in figure 10.1. We require the integral of the function between the ordinates $x = a$ and $x = b$. This is thus the area under the graph between those ordinates. The simplest approximation we can make is to draw a horizontal line through the mid-ordinate of that area at $x = x_1$. We have then replaced the area under the graph by a rectangular area which we can easily compute. What we have done is assume that the function between the ordinates can be approximated by the equation

$$y = A$$

where A is a constant, the constant being chosen to satisfy the mid-ordinate value. The area of the rectangle is the value of the function at x_1, i.e. y_1, multiplied by $(b - a)$. Thus

$$\int_a^b f(x)\, dx \approx (b - a)y_1 \qquad [1]$$

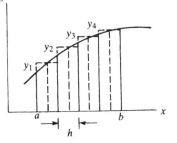

Fig. 10.2 The mid-ordinate rule

Replacing the area under a function by a single rectangle in this way gives an approximation which might be considered rather crude. However, we can improve it by subdividing the area under a function into more than one interval and then applying the mid-ordinate equation [1] to each separately. Figure 10.2 illustrates this with the division being into equal width strips of width h. The area under the function between the ordinates a and b is now given by

$$\int_a^b f(x)\, dx \approx y_1 h + y_2 h + \ldots + y_n h = h(y_1 + y_2 + \ldots + y_n)$$

and so

$$\int_a^b f(x)\, dx \approx (\text{width of interval})(\text{sum of ordinates}) \qquad [2]$$

for n strips with $nh = b - a$. This relationship is termed the *mid-ordinate rule*.

To illustrate the above, consider the evaluation of the integral

$$\int_0^2 (x^2 + 1)\, dx$$

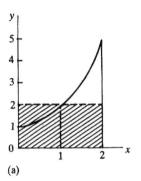

(a)

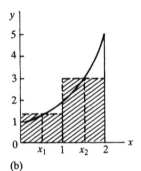

(b)

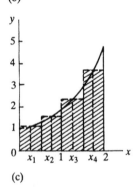

(c)

Fig. 10.3 The mid-ordinate rule with (a) one, (b) two, (c) four intervals

We can evaluate this integral analytically, obtaining the answer of 4.667. Now consider its evaluation by the mid-ordinate rule.

If we consider there to be only one interval between the ordinates $x = 0$ and $x = 2$, i.e. $h = 2$, then we have the situation shown in figure 10.3(a). The mid-ordinate is at $x = 1$ and the value of the function at this value is $y = (1^2 + 1) = 2$. The area of the rectangle is

$$\text{area} = 2 \times 2 = 4 \text{ square units}$$

If we have two intervals, i.e. $h = 1$, then the situation is as shown in figure 10.3(b). The mid-ordinates occur at $x_1 = 0.5$ and $x_2 = 1.5$. The values of y at these values are

$$y_1 = (0.5^2 + 1) = 1.25,$$
$$y_2 = (1.5^2 + 1) = 3.25$$

Hence, using equation [2], the shaded area is

$$\text{area} = 1(1.25 + 3.25) = 4.5 \text{ square units}$$

If we have four intervals, i.e. $h = 0.5$, then the situation is as shown in figure 10.3(c). The mid-ordinates occur at $x_1 = 0.25$, $x_2 = 0.75$, $x_3 = 1.25$ and $x_4 = 1.75$. The values of y at these values are

$$y_1 = (0.25^2 + 1) = 1.0625,$$
$$y_2 = (0.75^2 + 1) = 1.5625,$$
$$y_3 = (1.25^2 + 1) = 2.5625,$$
$$y_4 = (1.75^2 + 1) = 4.0625$$

Hence, using equation [2], the shaded area is

$$\text{area} = 0.5(1.0625 + 1.5625 + 2.5625 + 4.0625)$$

$$= 4.625 \text{ square units}$$

Thus the accuracy improves as the number of subdivisions increases.

Example

Use the mid-ordinate rule with (a) two intervals, (b) four intervals to evaluate the integral

$$\int_1^3 \frac{1}{\sqrt{x}}\, dx$$

(a) With two intervals we have $2h = 3 - 1$ and so $h = 1$. The mid-ordinates will thus occur at $x_1 = 1.5$ and $x_2 = 2.5$. At these values of x we have

$$y_1 = 1/\sqrt{1.5} = 0.816,$$
$$y_2 = 1/\sqrt{2.5} = 0.632$$

Hence

$$\int_1^3 \frac{1}{\sqrt{x}} \, dx \approx 1(0.816 + 0.632) = 1.448$$

(b) With four intervals we have $4h = 3 - 1$ and so $h = 0.5$. The mid-ordinates will thus occur at $x_1 = 1.25$, $x_2 = 1.75$, $x_3 = 2.25$ and $x_4 = 2.75$. At these values of x we have

$$y_1 = 1/\sqrt{1.25} = 0.894,$$
$$y_2 = 1/\sqrt{1.75} = 0.756,$$
$$y_3 = 1/\sqrt{2.25} = 0.667,$$
$$y_4 = 1/\sqrt{2.75} = 0.603$$

Hence

$$\int_1^3 \frac{1}{\sqrt{x}} \, dx \approx 0.5(0.894 + 0.756 + 0.667 + 0.603) = 1.460$$

This integral in this problem can easily be integrated analytically, giving the exact result of 1.464. Thus, increasing the number of intervals has improved the accuracy.

Review problems

1   Use the mid-ordinate rule with (a) two intervals, (b) four intervals to evaluate the integral

$$\int_{-1}^1 e^x \, dx$$

2   Use the mid-ordinate rule with (a) two intervals, (b) four intervals to evaluate the integral

$$\int_0^1 \frac{1}{x^2 + 1} \, dx$$

3   Use the mid-ordinate rule with ten subdivisions to evaluate the integral

$$\int_0^1 e^{-x^2}$$

10.3 The trapezium rule

Consider the function $y = f(x)$ and its integral between the values of $x = a$ and $x = b$. This is the area under the graph of the curve between the ordinates $x = a$ and $x = b$. Figure 10.4 shows the function. If we assume the approximation to the function joining the tops of two ordinates to be of the form

$$y = A + Bx$$

i.e. a straight line, then we can replace the function by the straight line shown in figure 10.4. Then we have

$$\int_a^b f(x)\,dx \approx \text{area of the trapezium}$$

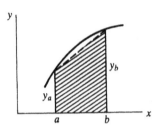

The area of a trapezium is given by

$$\text{area} = \tfrac{1}{2}(\text{sum of parallel sides})$$
$$\times (\text{perpendicular distance between them})$$

Fig. 10.4 The trapezium approximation

Thus, for the trapezium in figure 10.4,

$$\text{area} = \tfrac{1}{2}(y_a + y_b)(b - a)$$

Hence

$$\int_a^b f(x)\,dx \approx (b - a)\tfrac{1}{2}(y_a + y_b) \tag{3}$$

i.e. the width of the interval multiplied by half the sum of the first and last ordinates.

As suggested by figure 10.5, it looks more likely that we can obtain a better value for an integral if we subdivide the area beneath the curve into equal width strips and apply the trapezoidal equation [3] to each such area and then add the results. Thus, with the subdivision to give the four equal width strips shown in the figure,

$$\int_a^b f(x)\,dx \approx h\tfrac{1}{2}(y_a + y_1) + h\tfrac{1}{2}(y_1 + y_2) + h\tfrac{1}{2}(y_2 + y_3)$$
$$+ h\tfrac{1}{2}(y_3 + y_b)$$

$$\approx h\left[\tfrac{1}{2}(y_a + y_b) + y_1 + y_2 + y_3\right]$$

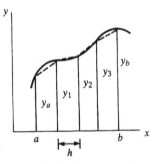

Fig. 10.5 The trapezium rule

with $4h = b - a$. In general we have, with n subdivisions,

$$\int_a^b f(x)\,dx \approx h\left[\tfrac{1}{2}(y_a + y_b) + \text{sum of other ordinates}\right] \tag{4}$$

with $nh = b - a$. Equation [4] is frequently referred to as the *trapezoidal rule*.

Example

Use the trapezoidal rule with four subdivisions to evaluate the integral

$$\int_1^3 \frac{1}{\sqrt{x}}\,dx$$

With four intervals we have $h = (3 - 1)/4 = 0.5$. The ordinates are at $x_a = 1$, $x_1 = 1.5$, $x_2 = 2.0$, $x_3 = 2.5$ and $x_b = 3.0$. The values of y at these ordinates are

$y_a = 1/\sqrt{1} = 1,$
$y_1 = 1/\sqrt{1.5} = 0.816,$
$y_2 = 1/\sqrt{2.0} = 0.707,$
$y_3 = 1/\sqrt{2.5} = 0.632,$
$y_b = 1/\sqrt{3.0} = 0.577$

Thus, using equation [4],

$$\int_1^3 \frac{1}{\sqrt{x}}\,dx \approx h\left[\tfrac{1}{2}(y_a + y_b) + \text{sum of other ordinates}\right]$$

$$\approx 0.5\left[\tfrac{1}{2}(1 + 0.577) + 0.816 + 0.707 + 0.632\right]$$

$$\approx 1.472$$

This integral in this problem can easily be integrated analytically, giving the exact result of 1.464. With a greater number of subdivisions a result closer to the exact result would be obtained. Note that this same integral was evaluated as an example earlier in this chapter using the mid-ordinate rule. You might like to compare the results.

Example

Using the trapezoidal rule with (a) four subdivisions, (b) eight subdivisions, evaluate the integral

$$\int_0^\pi \sin x\,dx$$

(a) With four subdivisions we have $h = \pi/4$. Thus we have ordinates at $x = 0$, $\pi/4$, $\pi/2$, $3\pi/4$ and π. The values of the function at these values of x are

$y_a = \sin 0,$
$y_1 = \sin \pi/4,$
$y_2 = \sin \pi/2,$
$y_3 = \sin 3\pi/4,$
$y_b = \sin \pi$

Thus, using equation [4],

$$\int_0^\pi \sin x \, dx \approx \tfrac{1}{4}\pi[\tfrac{1}{2}(\sin 0 + \sin \pi) + \sin \pi/4 + \sin \pi/2 + \sin 3\pi/4]$$

$$\approx \tfrac{1}{4}\pi[0 + 0.707 + 1 + 0.707]$$

$$\approx 1.896$$

(b) With eight subdivisions we have $h = \pi/8$. Thus we have ordinates at $x = 0$, $\pi/8$, $\pi/4$, $3\pi/8$, $\pi/2$, $5\pi/8$, $3\pi/4$, $7\pi/8$ and π. Thus the function has values of

$y_a = \sin 0$,
$y_1 = \sin \pi/8$,
$y_2 = \sin \pi/4$,
$y_3 = \sin 3\pi/8$,
$y_4 = \sin \pi/2$,
$y_5 = \sin 5\pi/8$
$y_6 = \sin 3\pi/4$
$y_7 = \sin 7\pi/8$
$y_b = \sin \pi$

Thus, using equation [4],

$$\int_0^\pi \sin x \, dx \approx \tfrac{1}{8}\pi[\tfrac{1}{2}(\sin 0 + \sin \pi) + \sin \pi/8 + \sin \pi/4 + \sin 3\pi/8$$
$$+ \sin \pi/2 + \sin 5\pi/8 + \sin 3\pi/4 + \sin 7\pi/8]$$

$$\approx \tfrac{1}{8}\pi[0 + 0.383 + 0.707 + 0.924 + 1 + 0.924$$
$$+ 0.707 + 0.383]$$

$$\approx 1.974$$

This integral in this problem can easily be integrated analytically, giving the exact result of 2. With a greater number of subdivisions a result closer to the exact result is obtained.

Review problems

4 Use the trapezoidal rule to evaluate the following integrals:

(a) $\int_1^2 \frac{1}{x} dx$ with four subdivisions,

(b) $\int_1^3 \frac{1}{x+2} dx$ with four subdivisions,

(c) $\int_0^{\pi/2} \frac{1}{\sin x + 1} dx$ with six subdivisions,

(d) $\int_0^2 x^3 \, dx$ with eight subdivisions,

(e) $\int_2^6 \sqrt{x^3 + 1} \, dx$ with ten divisions

5 Use the trapezoidal rule to determine the work done between displacements $x = 0.1$ m and $x = 0.5$ m, i.e.

$$\int_{0.1}^{0.5} F \, dx$$

when the forces F involved were measured as

Force in N	1.8	2.1	2.4	2.5	2.9
Displacement in m	0.1	0.2	0.3	0.4	0.5

6 Use the trapezoidal rule to determine the charge moved on to a capacitor plate between times $t = 0$ and $t = 1$ s, i.e.

$$\int_0^1 i \, dt$$

when the current i was recorded at a number of times in that interval to give the data

Current in A	0	0.1	0.3	0.7	1.1	1.1
Time in s	0	0.2	0.4	0.6	0.8	1.0

10.4 Simpson's rule

The mid-ordinate rule involved fitting the equation $y = A$ to a horizontal line through the mid-ordinate of each strip into which the area under the curve had been subdivided, the trapezoidal rule the fitting of the equation $y = A + Bx$ to a curve to join the top corners of the strips into which the area under the curve had been subdivided. Simpson's rule involves the fitting of the equation

$$y = A + Bx + Cx^2 \tag{5}$$

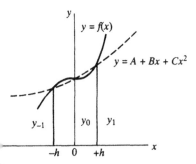

Fig. 10.6 The quadratic approximation

to a curve to join the top corners of the strips into which the area under the curve had been subdivided.

Consider the function $y = f(x)$ which is described by the graph shown in figure 10.6. The area beneath the curve, between the two indicated ordinates, is divided into two equal width strips. For simplicity the strips are shown as one on either side of the ordinate $x = 0$. The quadratic equation [5] is chosen with the constraint that it must join the top corners of each strip. It must therefore pass through the points $(-h, y_{-1})$, $(0, y_0)$ and (h, y_1).

Putting these values into equation [5] gives three simultaneous equations:

$$y_{-1} = A - Bh + Ch^2 \qquad [6]$$

$$y_0 = A + 0 + 0 \qquad [7]$$

$$y_1 = A + Bh + Ch^2 \qquad [8]$$

Adding equations [6] and [8] gives

$$y_{-1} + y_1 = 2A + 2Ch^2$$

Hence, when equation [7] is used to substitute for A, we have

$$y_{-1} + y_1 = 2y_0 + 2Ch^2 \qquad [9]$$

The area under the quadratic between the ordinates $x = -h$ and $x = h$ is

$$\text{area} = \int_{-h}^{h} (A + Bx + Cx^2)\, dx$$

$$= \left(Ah + \tfrac{1}{2}Bh^2 + \tfrac{1}{3}Ch^3\right) - \left(-Ah + \tfrac{1}{2}Bh^2 - \tfrac{1}{3}Ch^3\right)$$

$$= 2Ah + \tfrac{2}{3}Ch^3$$

Substituting for A by means of equation [7] and C by equation [9] gives

$$\text{area} = 2y_0 h + \tfrac{1}{3}h(y_{-1} + y_1 - 2y_0)$$

$$= \tfrac{1}{3}h(y_{-1} + 4y_0 + y_1) \qquad [10]$$

This area is the approximation for the area under the function $y = f(x)$ between the ordinates $x = -h$ and $x = h$. It is the *basic Simpson equation*.

The equation involves a double strip. For convenience, in the derivation the strips were positioned either side of the y-axis. This does not have to be the case (you might like to derive the equation for a more general situation); the pair of strips can be positioned either side of any x-value.

Now consider the function shown in figure 10.7 where the area under the function between the ordinates $y = a$ and $y = b$ is required. We can divide the area into a number of pairs of equal width strips, as shown in the figure. Then, if equation [10] is applied to each pair of strips we have for the estimate of the area under the function

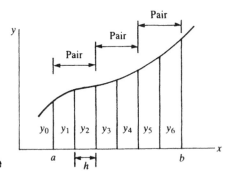

Fig. 10.7 Applying Simpson's rule

$$\text{area} \approx \tfrac{1}{3}h(y_0 + 4y_1 + y_2) + \tfrac{1}{3}h(y_2 + 4y_3 + y_4)$$
$$\cdot + \tfrac{1}{3}h(y_4 + 4y_5 + y_6)$$

$$\approx \tfrac{1}{3}h[(y_0 + y_6) + 4(y_1 + y_3 + y_5) + 2(y_2 + y_4)] \qquad [11]$$

This equation is *Simpson's rule*. We could have considered more strips, the only criterion being that we must have an even number so that they can form pairs. In general, we have

$$\int_a^b f(x)\, dx \approx \tfrac{1}{3}h[(\text{first} + \text{last ordinate})$$
$$+ 4(\text{sum of odd ordinates})$$
$$+ 2(\text{sum of even ordinates})] \qquad [12]$$

Note that in some texts the ordinates are not counted 0, 1, 2, etc. but 1, 2, 3, etc. and so what is termed odd in one case is even in the other.

Example

Use Simpson's rule with (a) four strips and (b) eight strips to evaluate the integral

$$\int_1^3 \frac{1}{\sqrt{x}}\, dx$$

(a) With four strips, each will have a width of $h = (3 - 2)/4 = 0.5$. Thus we have ordinates at $x_0 = 1.0$, $x_1 = 1.5$, $x_2 = 2.0$, $x_3 = 2.5$ and $x_4 = 3.0$. The values of the function at these ordinates are

$$y_0 = 1/\sqrt{1} = 1.0,$$
$$y_1 = 1/\sqrt{1.5} = 0.816,$$
$$y_2 = 1/\sqrt{2.0} = 0.707,$$
$$y_3 = 1/\sqrt{2.5} = 0.632,$$
$$y_4 = 1/\sqrt{3.0} = 0.577$$

Thus, using equation [12],

$$\int_1^3 \frac{1}{\sqrt{x}}\,dx \approx \tfrac{1}{3} \times 0.5[(1.0 + 0.577) + 4(0.816 + 0.632)$$
$$+ 2(0.707)]$$

$$\approx 1.464$$

(b) With eight strips, each will have a width of $h = (3 - 2)/8 = 0.25$. Thus we have ordinates at $x_0 = 1.0$, $x_1 = 1.25$, $x_2 = 1.5$, $x_3 = 1.75$, $x_4 = 2.0$, $x_5 = 2.25$, $x_6 = 2.5$, $x_7 = 2.75$ and $x_8 = 3.0$. The values of the function at these ordinates are

$$y_0 = 1/\sqrt{1} = 1.0,$$
$$y_1 = 1/\sqrt{1.25} = 0.894,$$
$$y_2 = 1/\sqrt{1.5} = 0.816,$$
$$y_3 = 1/\sqrt{1.75} = 0.756,$$
$$y_4 = 1/\sqrt{2.0} = 0.707,$$
$$y_5 = 1/\sqrt{2.25} = 0.667,$$
$$y_6 = 1/\sqrt{2.5} = 0.632,$$
$$y_7 = 1/\sqrt{2.75} = 0.603,$$
$$y_8 = 1/\sqrt{3.0} = 0.577$$

Thus, using equation [12],

$$\int_1^3 \frac{1}{\sqrt{x}}\,dx \approx \tfrac{1}{3} \times 0.25[(1.0 + 0.577)$$
$$+ 4(0.894 + 0.756 + 0.667 + 0.603)$$
$$+ 2(0.816 + 0.707 + 0.632)]$$

$$\approx 1.464$$

Thus to the accuracy of the third decimal place, the four strips are sufficient.

Example

Use Simpson's rule with four strips to evaluate the integral

$$\int_0^\pi \sin x\,dx$$

With four strips we have $h = \pi/4$. The ordinates thus occur at $x_0 = 0$, $x_1 = \pi/4$, $x_2 = \pi/2$, $x_3 = 3\pi/4$ and $x_4 = \pi$. At these ordinates we have

$$y_0 = \sin 0 = 0,$$
$$y_1 = \sin \pi/4 = 0.707,$$
$$y_2 = \sin \pi/2 = 1,$$

$$y_3 = \sin 3\pi/4 = 0.707,$$
$$y_4 = \sin \pi = 0$$

Thus, using equation [12],

$$\int_0^\pi \sin x \, dx \approx \tfrac{1}{3} \times \tfrac{1}{4}\pi[(0 + 0) + 4(0.707 + 0.707) + 2(1)]$$

$$\approx 2.004$$

Review problems

7 Use Simpson's rule to evaluate the following integrals:

(a) $\int_1^3 \dfrac{1}{2+x} \, dx$ with 4 strips,

(b) $\int_0^1 e^{2x} \, dx$ with 4 strips,

(c) $\int_0^1 (1+x^2)^{3/2} \, dx$ with 4 strips,

(d) $\int_0^1 \cos x^2 \, dx$ with 4 strips,

(e) $\int_0^1 \dfrac{1}{x^2+1} \, dx$ with 4 strips,

(f) $\int_0^{\pi/2} \sqrt{\sin x} \, dx$ with 6 strips,

(g) $\int_1^3 \sqrt{x^4+1} \, dx$ with 6 strips,

(h) $\int_1^3 \ln \sqrt{x} \, dx$ with 10 strips

8 The following results were obtained from measurements of the current i in a circuit as a function of time t. Determine the charge q moved in the time interval $t = 0$ to $t = 2$ s if

$$q = \int_0^2 i \, dt$$

i in mA	1.000	1.649	2.718	4.482	7.389
t in s	0	0.5	1.0	1.5	2.0

9 The following results were obtained from measurements of the velocity v of a car as a function of time t. Determine the distance d travelled by the car in the time $t = 0$ to $t = 6$ s, i.e. evaluate

$$d = \int_0^6 v \, dt$$

v in m/s	0	0.2	0.6	1.4	2.2	3.0	3.8
t in s	0	1	2	3	4	5	6

10 A river has a width of 12 m. Depth measurements every 2 m across the river gives depths of

Depth in m	0	1.8	3.0	4.2	2.1	1.1	0
Distance in m across river	0	2	4	6	8	10	12

Use Simpson's rule to obtain an estimate of the cross-sectional area of the river.

10.5 Errors

Consider the evaluation of the integral

$$\int_1^3 \frac{1}{\sqrt{x}} \, dx$$

This integral was an example used in the earlier sections of this chapter for numerical integration by the mid-ordinate rule, the trapezoidal rule and Simpson's rule. Analytically, the value of the integral is 1.464. The error of a value obtained by numerical integration is

error = true value of integral − calculated value   [13]

The following are the errors obtained with the above integral when numerically integrated by different methods and with different values of h.
    With the *mid-ordinate rule* the results are

2 intervals, $h = 1$	1.448	error 0.016
4 intervals, $h = 0.5$	1.460	error 0.004
8 intervals, $h = 0.25$	1.463	error 0.001

Doubling the number of intervals, i.e. halving h, reduces the error by a factor of 4. The error can be considered to occur mainly as a result of limiting the number of terms used in the polynomial used to replace the function. As a consequence we have

error $\propto h^2$ [14]

With the *trapezoidal rule* the results obtained were

2 subdivisions, $h = 1$ 1.496   error −0.032

4 subdivisions, $h = 0.5$ 1.472   error −0.008

8 subdivisions, $h = 0.25$ 1.466   error −0.002

Doubling the number of intervals, i.e. halving h, reduces the error by a factor of 4. The error can be considered to occur mainly as a result of limiting the number of terms used in the polynomial used to replace the function. As a consequence we have

error $\propto h^2$ [15]

With *Simpson's rule* the results obtained were (true answer 1.464 102 and more decimal places included with these than with the previous calculations)

2 strips, $h = 1$ 1.468 592   error −0.007 576

4 strips, $h = 0.5$ 1.464 562   error −0.000 460

8 strips, $h = 0.25$ 1.464 137   error −0.000 036

The convergence to the true result as h increases is far faster with Simpson's rule than with the previous two methods. Doubling the number of strips considered, i.e. reducing h by a factor of 2, reduces the error by a factor of about 16, i.e. h^4.

error $\propto h^4$ [16]

Because Simpson's rule converges so rapidly to the true result, the number of strips required to obtain a given accuracy is much lower than the other methods considered and so fewer calculations are involved. For this reason, Simpson's rule is generally the preferred method when compared with the mid-ordinate rule and the trapezoidal rule. There are, however, situations when Simpson's rule, and the trapezium rule, will not give an answer. This is when the initial ordinate goes off to infinity, i.e. the integral of $1/x$ between the ordinates 0 and 1. The mid-ordinate rule, however, can still be applied since, unlike the other two methods, it does not involve the initial ordinate.

Review problems

11 Estimate the value of the integral

$$\int_0^1 \frac{1}{x^2 + 1}\, dx$$

to four decimal places using (a) the trapezium rule with five ordinates, (b) Simpson's rule with four strips. Evaluate the integral analytically and hence determine the accuracy of the numerically obtained results.

Hint: for the analytical integration, you can try the substitution $x = \tan \theta$.

12 Estimate the value of the integral

$$\int_1^4 \ln x \, dx$$

to four decimal places using (a) the trapezium rule with seven ordinates, (b) Simpson's rule with six strips. Evaluate the integral analytically and hence determine the accuracy of the numerically obtained results.

Hint: see the example in section 6.3 for the analytical integration.

10.6 Romberg integration

There is a method that can be used to improve on the accuracy of the trapezoidal rule. It is known as *Romberg integration*.

We have (equation [13])

true value = calculated value + error

and since the error for the trapezoidal rule is proportional to h^2 (equation [15]) we can write

$$\text{true value} = \text{calculated value} + Ch^2 \qquad [17]$$

where C is some constant. Suppose we calculate the value, y_{2h}, with some value of $2h$.

$$\text{true value} = y_{2h} + C(2h)^2 \qquad [18]$$

Now suppose we calculate the value with the value of the interval reduced to half its value, i.e. h. This will give us a more accurate value. Then

$$\text{true value} = y_h + Ch^2 \qquad [19]$$

We can eliminate the Ch^2 term from the simultaneous equations [18] and [19] by subtracting equation 4 × [19] from [18].

$$-3 \, (\text{true value}) = y_{2h} - 4y_h$$

We can write this as

$$\text{true value} = y_h + \tfrac{1}{3}(y_h - y_{2h}) \qquad [20]$$

We have thus eliminated the error which is proportional to h^2. The error is, however, only approximately proportional to h^2 and a better approximation for the error is to consider that it consists of h^2, h^4, h^6, etc. terms. Thus one application of the Romberg method results in the removal of the h^2 errors. If we repeat the method with a yet further halving of h we can eliminate the h^4 error. The method can be repeated until we have a result of the required accuracy.

Since we may not know the true value of the integral, equation [20] can be written as

$$\begin{aligned}
\text{improved value} \\
= \text{more accurate value} \\
+ \left(\frac{1}{2^n - 1} \right) (\text{more accurate} - \text{less accurate values}) \quad [21]
\end{aligned}$$

where n is the power of h that applies to the error term being eliminated. The more accurate value is the one obtained with the smallest value of h. This method is widely used in computer programs used for carrying out integration.

Example

Use the Romberg method to eliminate the h^2 error with the following results obtained by the use of the trapezium rule for the integral

$$\int_1^3 \frac{1}{\sqrt{x}} \, dx$$

2 subdivisions, $h = 1$	1.496
4 subdivisions, $h = 0.5$	1.472

Applying equation [20],

$$\text{improved value} = 1.472 + \tfrac{1}{3}(1.472 - 1.496) = 1.464$$

Review problems

13 Use the Romberg method to eliminate the h^2 error with the following results obtained by the use of the trapezium rule for the integral

$$\int_0^\pi \sin x \, dx$$

$$h = \pi/4 \qquad 1.896$$
$$h = \pi/8 \qquad 1.974$$

14 Use the Romberg method to eliminate the h^2 error with the following results obtained by the use of the trapezium rule for the integral

$$\int_0^1 e^{-x^2}\, dx$$

$$h = 0.1 \qquad 0.747\ 131$$
$$h = 0.05 \qquad 0.746\ 901$$

Further problems

15 Use the mid-ordinate rule to evaluate the following integrals:

(a) $\int_1^2 \dfrac{1}{x^2 + 1}\, dx$ with 8 subdivisions,

(b) $\int_1^5 (x^2 - 2x + 2)\, dx$ with 8 subdivisions,

(c) $\int_0^{\pi/3} \sqrt{\cos^3 x}\, dx$ with 6 subdivisions

16 Use the trapezoidal rule to evaluate the following integrals:

(a) $\int_0^2 x^2\, dx$ with 4 subdivisions,

(b) $\int_1^2 \dfrac{1}{x^2}\, dx$ with 4 subdivisions,

(c) $\int_0^1 \sqrt{x^3 + 2}\, dx$ with 4 subdivisions,

(d) $\int_0^{\pi/3} \sqrt{\sin x}\, dx$ with 6 subdivisions,

(e) $\int_1^3 \ln \sqrt{x}\, dx$ with 10 subdivisions

17 Use the trapezoidal rule to determine the value of the integral

$$\int_1^5 i\, dt$$

given the following data:

i in mA	0.41	0.45	1.25	2.85	6.05
t in s	1	2	3	4	5

18 Use Simpson's rule to evaluate the following integrals:

(a) $\int_0^2 e^x \, dx$ with 4 strips,

(b) $\int_0^1 \sqrt{x - x^2} \, dx$ with 4 strips,

(c) $\int_0^1 \frac{2}{1 + x^2} \, dx$ with 4 strips,

(d) $\int_0^{\pi/4} x \tan x \, dx$ with 4 strips,

(e) $\int_1^4 \frac{1}{x^3} \, dx$ with 6 strips,

(f) $\int_1^{1.6} \frac{\sin 2x}{x} \, dx$ with 6 strips,

(g) $\int_{-1}^1 e^{x^2} \, dx$ with 8 strips

19 A graph is plotted with the following values. Determine the area under the graph between the ordinates $x = 0$ and $x = 4$.

y	0.4	0.9	1.2	1.5	1.6
x	0	1	2	3	4

20 Evaluate the integral

$$\int_0^2 x^3 \, dx$$

to four decimal places by means of (a) the trapezoidal rule with five ordinates, (b) Simpson's rule with four strips. Evaluate the integral analytically and so determine the errors associated with each numerical method.

21 Evaluate the integral

$$\int_1^2 (3x^2 - 1) \, dx$$

to four decimal places by means of (a) the trapezoidal rule with five ordinates, (b) Simpson's rule with four strips. Evaluate the integral analytically and so determine the errors associated with each numerical method.

22 Use the Romberg method to eliminate the h^2 error with the trapezium rule used with $h = 1$ and $h = 0.5$ and obtain an improved value for the integral

$$\int_1^4 \ln x \, dx$$

23 Use the Romberg method to eliminate the h^2 error with the trapezium rule used with $h = \pi/8$ and $h = \pi/16$ and obtain an improved value for the integral

$$\int_{\pi/4}^{\pi/2} \cos x \, dx$$

24 Use the Romberg method to eliminate the h^2 error with the trapezium rule used with $h = 0.5$ and $h = 0.25$ and obtain an improved value for the integral

$$\int_0^1 (1 - x^3) \, dx$$

25 Use the Romberg method to eliminate the h^2 error with the trapezium rule used with $h = 1$ and $h = 0.5$ and obtain an improved value for the integral

$$\int_1^3 \frac{1}{x+2} \, dx$$

Appendix
Supporting mathematics

The following is a brief review of some of the mathematics assumed in this book.

Functions

There are many relationships between two variables in which the value of one of the variables depends on the value of the other. Thus, for example, the area A of a circle depends on its radius r, being given by $A = \pi r^2$. If two quantities are related, like the area and the radius, so that the value of one of them is uniquely determined when the other is known, then we say that there is a *functional relationship* between them. We can think of there being a black box with an input and an output. The input is one of the quantities and the output is the other quantity, the black box representing the function which operates on the input and changes it into the output. Thus we have an input of the radius into the box and an output of the area. With such a relationship there is only one value of the area which corresponds to a particular value of the radius. The area is said to be a function of the radius. We can express this as $A = f(r)$. This assumes that we put into the relationship a value of r and obtain as output a value of A.

The symbol $f(x)$ is generally used to denote a function of x. Although we generally use f as a convenient symbol, where we have a number of different functions other letters might be used, e.g. $y = g(x)$. If $y = f(x)$, then $f(a)$ denotes the value of y when we have $x = a$.

The inverse function

There are situations where we might want to put into the relationship between area and radius a value of the area and obtain the radius, i.e. do the above calculation in reverse. We can think of this as reversing the direction used in $A = f(r)$, i.e. we have taken the output of this functional relationship and found the input that could have caused it. We write this as $r = f^{-1}(A)$ and call it an *inverse function*.

Examples of functions

A *polynomial function* is one that can be expressed in the form

$$y = f(x) = A + Bx + Cx^2 + Dx^3 + ...$$

where A, B, C, D, etc. are constants.

A *trigonometric function* is one that can be expressed in terms of sines, cosines, tangents, secants, cosecants or cotangents of the variable, e.g.

$$y = f(x) = \sin x$$

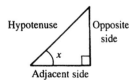

Hypotenuse / Opposite side

x

Adjacent side

Fig. Ap.1 Right-angled triangle

These trigonometric functions are defined in terms of the sides of a right-angled triangle (figure Ap.1):

$$\sin x = \frac{\text{opposite side}}{\text{hypotenuse}}$$

$$\cos x = \frac{\text{adjacent side}}{\text{hypotenuse}}$$

$$\tan x = \frac{\sin x}{\cos x}$$

$$\csc x = \frac{1}{\sin x}$$

$$\sec x = \frac{1}{\cos x}$$

$$\cot x = \frac{1}{\tan x}$$

Note that these relationships have many values of x for a particular value of y, e.g. $1 = \sin \pi/2 = \sin 5\pi/2 = \sin 9\pi/2$, etc. The function can be made a one-to-one function by restricting the values that x is allowed to take. Without this there will be ambiguity; in terms of the black box representation of the function we will not know what output to expect from a particular input.

The inverse of a trigonometric function, e.g. the inverse of $y = \sin x$, is written in the form

$$x = \sin^{-1}y$$

and is read as: x is the angle whose sine is y. The -1 in $\sin^{-1}x$ is *not* a power. It does not mean $1/\sin x$. For this function to be a one-to-one function the values permitted for x have to be restricted.

An *exponential function* is one that can be expressed in the form

$$y = f(x) = e^x$$

where e is a special number which is approximately equal to 2.718. As x increases positively then e^x increases, i.e. as $x \to \infty$ then $e^x \to \infty$. As x increases negatively then e^x approaches zero, i.e. as $x \to -\infty$ then $e^x \to 0$. Note that the exponential function never has a negative value.

The inverse of the exponential function is called the *natural logarithm function*.

$$x = \ln y$$

Note that because the exponential function can never have a negative value then we cannot talk of the natural logarithm of a negative quantity.

Associated with the exponential function is a family of functions called *hyperbolic functions*, e.g.

$$y = f(x) = \cosh x$$

These hyperbolic functions are defined as (see figures 1.8, 1.9 and 1.10):

$$\cosh x = \tfrac{1}{2}(e^x + e^{-x})$$

$$\sinh x = \tfrac{1}{2}(e^x - e^{-x})$$

$$\tanh x = \frac{\sinh x}{\cosh x}$$

$$\operatorname{sech} x = \frac{1}{\cosh x}$$

$$\operatorname{cosech} x = \frac{1}{\sinh x}$$

$$\coth x = \frac{1}{\tanh x}$$

The inverse of a hyperbolic function is written in the form

$$x = \sinh^{-1} y$$

y is the number whose sinh is x. In order that we have only a one-to-one relationship, the values have to be restricted for the inverse cosh and tanh functions. See later in this appendix for further discussion of the inverse hyperbolic functions.

Trigonometric relationships

If we apply the Pythagoras theorem to the right-angled triangle in figure Ap.1, then

$$(\text{hypotenuse})^2 = (\text{opposite side})^2 + (\text{adjacent side})^2$$

Dividing throughout by $(\text{hypotenuse})^2$ gives

$$1 = \sin^2 x + \cos^2 x$$

By dividing this equation throughout by $\cos^2 x$ we can arrive at another relationship

$$\sec^2 x = 1 + \tan^2 x$$

If we had divided by $\sin^2 x$ we would have obtained

$$\operatorname{cosec}^2 x = \cot^2 x + 1$$

It is often useful to express the trigonometric function of compound angles such as $A + B$ or $A - B$ in terms of the trigonometric functions of A and of B. Consider the two right-angled triangles shown in figure Ap.2.

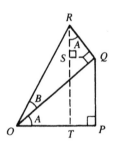

Fig. Ap.2 Two right-angled triangles

$$\sin(A+B) = \frac{TR}{OR} = \frac{TS + SR}{OR} = \frac{PQ + SR}{OR}$$

$$= \frac{PQ}{OQ} \times \frac{OQ}{OR} + \frac{SR}{QR} \times \frac{QR}{OR}$$

Hence

$$\sin (A + B) = \sin A \cos B + \cos A \sin B$$

In a similar way we can derive:

$$\sin (A - B) = \sin A \cos B - \cos A \sin B$$

$$\cos (A + B) = \cos A \cos B - \sin A \sin B$$

$$\cos (A - B) = \cos A \cos B + \sin A \sin B$$

$$\tan (A + B) = \frac{\tan A + \tan B}{1 - \tan A \tan B}$$

$$\tan (A - B) = \frac{\tan A - \tan B}{1 + \tan A \tan B}$$

But putting $A = B$ we can obtain the following double angle equations:

$$\sin 2A = 2 \sin A \cos A$$

$$\cos 2A = \cos^2 A - \sin^2 A = 2 \cos^2 A - 1 = 1 - 2 \sin^2 A$$

$$\tan 2A = \frac{2 \tan A}{1 - \tan^2 A}$$

Adding and subtracting the equations for the sums and differences of angles leads to the following equations in which products of angles are converted into sums or differences:

$$\sin A \cos B = \tfrac{1}{2}[\sin (A + B) + \sin (A - B)]$$

$$\cos A \sin B = \tfrac{1}{2}[\sin (A + B) - \sin (A - B)]$$

$$\cos A \cos B = \tfrac{1}{2}[\cos (A + B) + \cos (A - B)]$$

$$\sin A \sin B = -\tfrac{1}{2}[\cos (A + B) - \cos (A - B)]$$

Similarly we can obtain equations converting sums or differences into products:

$$\sin A + \sin B = 2 \sin \left(\frac{A + B}{2}\right) \cos \left(\frac{A - B}{2}\right)$$

$$\sin A - \sin B = 2 \cos \left(\frac{A + B}{2}\right) \sin \left(\frac{A - B}{2}\right)$$

$$\cos A + \cos B = 2 \cos \left(\frac{A + B}{2}\right) \cos \left(\frac{A - B}{2}\right)$$

$$\cos A - \cos B = -2 \sin \left(\frac{A + B}{2}\right) \sin \left(\frac{A - B}{2}\right)$$

Hyperbolic functions

Using the definitions of $\cosh x$ and $\sinh x$ given earlier in this appendix, we can obtain

$$\cosh x + \sinh x = e^x$$

$$\cosh x - \sinh x = e^{-x}$$

Hence

$$(\cosh x + \sinh x)(\cosh x - \sinh x) = e^x e^{-x} = 1$$

Thus

$$\cosh^2 x - \sinh^2 x = 1$$

From this we can derive

$$1 - \tanh^2 x = \text{sech}^2 x$$

$$\coth^2 x - 1 = \text{cosech}^2 x$$

Similarly we can show, for compound angle addition and subtraction:

$$\sinh (A + B) = \sinh A \cos B + \cosh A \sinh B$$

$$\sinh (A - B) = \sinh A \cosh B - \cosh A \sinh B$$

$$\cosh (A + B) = \cosh A \cosh B + \sinh A \sinh B$$

$$\cosh (A - B) = \cosh A \cosh B - \sinh A \sinh B$$

$$\tanh (A + B) = \frac{\tanh A + \tanh B}{1 + \tanh A \tanh B}$$

$$\tanh (A - B) = \frac{\tanh A - \tanh B}{1 - \tanh A \tanh B}$$

When $A = B$ the above give the double angle equations

$$\sinh 2A = 2 \sinh A \cosh A$$

$$\cosh 2A = \cosh^2 A + \sinh^2 A = 2 \cosh^2 A - 1 = 1 + 2 \sinh^2 A$$

$$\tanh 2A = \frac{2 \tanh A}{1 + \tanh^2 A}$$

Rearrangement of equations from the above give products into sums or difference equations:

$$\sinh A \cosh B = \tfrac{1}{2}[\sinh (A + B) + \sinh (A - B)]$$

$$\cosh A \sinh B = \tfrac{1}{2}[\sinh (A + B) - \sinh (A - B)]$$

$$\cosh A \cosh B = \tfrac{1}{2}[\cosh (A + B) + \cosh (A - B)]$$

$$\sinh A \sinh B = -\tfrac{1}{2}[\cosh (A + B) - \cosh (A - B)]$$

and sums or differences into product equations:

$$\sinh A + \sinh B = 2 \sinh \left(\frac{A+B}{2}\right) \cosh \left(\frac{A-B}{2}\right)$$

$$\sinh A - \sinh B = 2 \cosh \left(\frac{A+B}{2}\right) \sinh \left(\frac{A-B}{2}\right)$$

$$\cosh A + \cosh B = 2 \cosh \left(\frac{A+B}{2}\right) \cosh \left(\frac{A-B}{2}\right)$$

$$\cosh A - \cosh B = 2 \sinh \left(\frac{A+B}{2}\right) \sinh \left(\frac{A-B}{2}\right)$$

Inverse hyperbolic functions

The inverse hyperbolic functions can be expressed in terms of logarithms. Thus, for example, if we have $y = \sinh^{-1}x$ then this implies $x = \sinh y$. Thus

$$x = \sinh y = \tfrac{1}{2}(e^y - e^{-y}) = \tfrac{1}{2}\left(e^y - \frac{1}{e^y}\right)$$

Hence

$$(e^y)^2 - 2x(e^y) - 1 = 0$$

This is a quadratic equation of the form $ax^2 + bx + c = 0$ and will have the roots

$$e^y = \frac{2x \pm \sqrt{4x^2 + 4}}{2} = x \pm \sqrt{x^2 + 1}$$

Since e^y must be greater than 1 we can discount the negative value of the root and so

$$e^y = x + \sqrt{x^2 + 1}$$

Thus

$$y = \ln\left(x + \sqrt{x^2 + 1}\right)$$

and so

$$\sinh^{-1}x = \ln\left(x + \sqrt{x^2 + 1}\right)$$

Similar equations can be developed for the inverse cosh and tanh functions, namely

$$\cosh^{-1}x = \ln\left(x + \sqrt{x^2 - 1}\right)$$

$$\tanh^{-1}x = \tfrac{1}{2}\ln\left(\frac{1+x}{1-x}\right)$$

Answers

1 (a) 4, (b) 10

2 0

3 (a) 5, (b) 10, (c) 1

4 (a) 0, (b) negative, (c) positive

5 (a) +1, (b) –2/3

6 (a) 0, (b) 2

7 (a) $x = 0$, π, 2π, (b) $x = \pi/2$, $3\pi/2$

8 (a) 1.7 m/s, (b) 2 m/s

9 (a) 3, 7, (b) 2, 4, (c) 2, 2

10 (a) $x = 0$, (b) $x = 0$, (c) $x = 1$

11 (a) $3x^2$, (b) $6x^5$, (c) $-2x^{-3}$, (d) $\frac{3}{2}x^{1/2}$, (e) $\frac{1}{2}x^{-1/2}$, (f) $4x$,

(g) $-\frac{3}{2}x^{-3/2}$, (h) 0

12 $8t$

13 6

14 125.7 mm²/mm

15 (a) $-\sin x$, (b) $5\cos 5x$, (c) $-4\sin 4x$, (d) $4\sec^2 4x$

16 $v = -20\sin 20t$

17 (a) $2e^{2x}$, (b) $-4e^{-4x}$, (c) $\frac{3}{2}e^{3x/2}$

18 $-\lambda N_0 e^{-\lambda t} = -\lambda N$

19 $2/x$

20 (a) $2\sinh 2x$, (b) $5\cosh 5x$, (c) $7\,\text{sech}^2 7x$

21 (a) 4, (b) $-\sin x$, (c) $4e^{2x}$, (d) $-1/x^2$

22 (a) $i = -\dfrac{V}{R}e^{-t/RC}$, (b) $\dfrac{di}{dt} = \dfrac{V}{R^2 C}e^{-t/RC}$

223

23 (a) $\dfrac{1}{2\sqrt{x}}$, (b) $-\dfrac{2}{x^2}$

24 (a) $6x$, (b) $-2x^{-3}$, (c) $\tfrac{5}{3}x^{2/3}$, (d) $6x$, (e) $x^{-1/2}$, (f) $-\tfrac{15}{2}x^{-5/2}$, (g) 0,

    (h) $4\cos 4x$, (i) $-2\sin 2x$, (j) e^x, (k) $2\,e^{2x}$, (l) $-\tfrac{2}{3}e^{-2x/3}$,

    (m) $1/x$, (n) $5\sec^2 5x$, (o) $5\cos 5x$, (p) $3\cosh 3x$,

    (q) $2\operatorname{sech}^2 2x$

25 $v = at$

26 (a) 0, (b) -2, (c) 0

27 40 mm^2/mm

28 $I\omega$

29 CV

30 $-p_0 c\,e^{-ch}$

31 (a) $36x^2$, (b) $4\,e^{2x}$, (c) $-4\cos 2x$, (d) $\tfrac{3}{4}x^{-5x/2}$

32 $\dfrac{dT}{dt} = -v\dfrac{d^2 m}{dt^2}$

33 $a = \dfrac{d^2 x}{dt^2} = -\omega^2 x$

Chapter 2

1 (a) $3 + 8x$, (b) $1 + 4\cos 4x$, (c) $2x - 3x^2$, (d) $2\,e^{2x} - 3\,e^{3x}$,

    (e) $0 + 3\,e^{-3x}$, (f) $-2\sin 2x - 2\cos 2x$

2 $a + 2bt$

3 $\dfrac{RI}{L}e^{-Rt/L}$

4 (a) $2 + 12x^3$, (b) $10\cos 2x$, (c) $10x^4 + 6\cos 2x$

5 $10 + 40t$

6 (a) $-4x^2\sin 4x + 2x\cos 4x$, (b) $-2x\,e^{-2x} + e^{-2x}$,

    (c) $-2\sin x \sin 2x + \cos x \cos 2x$, (d) $(2 + x^2)\cos x + 2x\sin x$,

    (e) $-3e^{2x}\sin 3x + 2e^{2x}\cos 3x$, (f) $5x^4 + 6x^2 + 2x$, (g) $2x + 4$,

    (h) $4(x^2 + x)^3(2x + 1)$

7 $-200\,e^{-3t}\sin 100t + 6\,e^{-3t}\cos 100t$

8 80 m

9 3.25

10 (a) $\dfrac{3x^2 + 4x}{(3x+2)^2}$, (b) $\dfrac{2 - 2x^2}{(x^2 + 1)^2}$, (c) $-\dfrac{16x}{(x^2 - 4)^2}$, (d) $\dfrac{2 + \sqrt{x}}{2(1 + \sqrt{x})^2}$,

    (e) $\dfrac{e^{4x}(\sin x + 4x\sin x - x\cos x)}{\sin^2 x}$, (f) $\dfrac{4(x-1)}{(x^2 + 1)^2}$, (g) $\dfrac{6(2x - 1)}{x^2(x - 1)^2}$

11 (a) $\sec^2\theta$, (b) $-\operatorname{cosec}^2\theta$, (c) $\sec\theta\tan\theta$

12 $\dfrac{x \sin 2x - x^2}{\sin^2 x}$

13 (a) $5(x + 2)^4$, (b) $2 \cos(2x + 1)$, (c) $2 \sin x \cos x = \sin 2x$,

(d) $\dfrac{3x^2}{2\sqrt{1 + x^3}}$, (e) $\dfrac{x}{\sqrt{1 + x^2}} \cos \sqrt{1 + x^2}$, (f) $4 e^{\sin x} \cos x$,

(g) $\dfrac{3(x - 2)^2(-2x^2 + 8x + 1)}{(2x + 1)^4}$, (h) $\dfrac{3(4x - 3x^2)}{(x^3 - 2x^2 - 1)^4}$,

(i) $3\left(x + \dfrac{1}{x}\right)^2\left(1 - \dfrac{1}{x^2}\right)$, (j) $-\dfrac{2 \sin 2x}{(1 - \cos 2x)^2}$

14 $0.0144 \text{ m}^3/\text{s}$

15 $471 \text{ mm}^2/\text{min}$

16 0.88 rad/s

17 (a) $\dfrac{2x}{3y^2 + 2y - 3}$, (b) $\dfrac{4x + 1}{6y}$, (c) $-\dfrac{y^3 + 2x}{3y^2 x + 4y^2}$, (d) $\dfrac{1 - 4x}{3y^2}$,

(e) $\dfrac{y - 2x}{2y - x}$, (f) $\dfrac{2(1 + 2x)}{y}$

18 (a) $\dfrac{1}{2y + 5}$, (b) $\dfrac{1}{3y^2 + 4y}$

19 (a) $\pi/3$, (b) $\pi/2$, (c) $\pi/3$

20 (a) $\dfrac{3}{1 + 9x^2}$, (b) $-\dfrac{2}{\sqrt{1 - 4x^2}}$, (c) $\dfrac{1}{\sqrt{25 - x^2}}$, (d) $\dfrac{3}{2 + 6x + 9x^2}$,

(e) $\dfrac{4x}{\sqrt{1 - 4x^4}}$, (f) $2t \cos^{-1}(t - 1) - \dfrac{x^2}{\sqrt{2x - x^2}}$, (g) $\dfrac{1}{x\sqrt{x - 1}}$,

(h) $-\dfrac{2x}{1 + x^4}$

21 (a) $\dfrac{3}{\sqrt{9x^2 - 1}}$, (b) $\sec x$, (c) $\dfrac{1}{\sqrt{1 + x^2}}$, (d) $\dfrac{3}{1 - 9x^2}$

22 $\dfrac{2t}{3} = \dfrac{2y}{3x} = \tfrac{2}{3}x^{-1/3}$

23 $2t + 1$

24 $-\dfrac{3}{2 \tan \theta}$, -0.87

25 $1/t$, $1/2$

26 (a) $2x \sin x + x^2 \cos x$, (b) $3x \, 2^{2x+1} \ln 2$,

(c) $2^x \tan x(\ln 2 + \sec x \csc x)$, (d) $-\dfrac{1}{\tan x}\left(2x + \dfrac{2 - x^2}{\sin x \cos x}\right)$,

(e) $e^{2x} \sin^3 x \cos^2 x \,(2 + 3 \cot x - 2 \tan x)$,

(f) $[(3 - 2x) \ln x + 1] \, x^2 e^{-2x}$

27 (a) $3 + 2 \cos 2x$, (b) $0 + 2x + 9x^2 + 4x^3$, (c) $2 \cos x - 3 \sin 3x$,

(d) $-4x^2 \sin 4x + 2x \cos 4x$, (e) $(2 + 3x)\cos x + 3 \sin x$,

(f) $4e^x \cos 2x + 2e^x \sin 2x$, (g) $4(2 + e^{-x})\cos 4x - 4e^{-x} \cos 4x$,

(h) $8x^3 + 3x^2 + 4$, (i) $3(x + 2)^2$, (j) $2(x^2 + 2x)^3(x + 1)$,

(k) $\cos x - x \sin x$, (l) $5 \sin x + (5x - 1)\cos x$,

(m) $\dfrac{2x\cos x + (x^2 - 1)\sin x}{\cos^2 x}$, (n) $-\dfrac{5}{(3x+2)^2}$, (o) $\dfrac{-5x^2 + 15}{(x^2 + 3)^2}$,

(p) $-2\,\mathrm{cosec}^2 2x$, (q) $\dfrac{x^2\cos x - x\cos x - \sin x}{(x-1)^2}$,

(r) $4(6x - 3)(2x^3 - 3x)^3$, (s) $-5\cos^4 x \sin x$, (t) $4\,e^{2x+1}$,

(u) $2\cos(2x + 3)$, (v) $-\dfrac{21}{(1+3x)^8}$, (w) $-\dfrac{4x}{2y}$, (x) $\dfrac{6x-4}{6y^2}$,

(y) $\dfrac{-3y - 4x}{3y^2 + 3x}$, (z) $\dfrac{5}{\sqrt{1 - 25x^2}}$, (aa) $-\dfrac{4}{\sqrt{1 - 16x^2}}$,

(ab) $\dfrac{2}{1 + 4x^2}$, (ac) $-\dfrac{1}{2 - 2x + x^2}$, (ad) $\dfrac{1}{2\sqrt{x - x^2}}$,

(ae) $\dfrac{2x^2}{\sqrt{1 - x^4}} + \sin^{-1}x^2$, (af) $\dfrac{1}{x^2}\left(\dfrac{2x}{\sqrt{1 - 4x^2}} - \sin^{-1}2x\right)$,

(ag) $\dfrac{2\cos x}{\sqrt{1 - 4\sin^2 x}}$, (ah) $-\dfrac{1}{x\sqrt{x^2 - 1}}$, (ai) $\dfrac{4}{\sqrt{16x^2 - 1}}$,

(aj) $-\dfrac{1}{(2x - 1)\sqrt{x(1 - x)}}$, (ak) $\dfrac{x}{\sqrt{x^2 + 1}} + \sinh^{-1}x$,

(al) $\dfrac{(x + 2)(x + 1)}{(x + 3)^2}\left(\dfrac{1}{x + 2} + \dfrac{1}{x + 1} + \dfrac{2}{x + 3}\right)$,

(am) $4x^{1/2}\cos 2x + x^{-1/2}\sin 2x$, (ao) $e^{2x}(\sec^2 x + 2\tan x)$

28 $20 - 8t$

29 $-500\cos 5t$

30 $50/\sqrt{T}$

31 $L_0(a + 200b)$

32 0.050 m/s

33 $-\dfrac{f_s}{c(1 + v/c)^2}$

34 1.4×10^7 mm³/min

35 0.5 rad/s

36 $\dfrac{9 + 6t}{2x}$

37 As given in the problem

38 As given in the problem

39 $5(t + 1)$

40 $-\dfrac{5}{2\tan\theta}$, -4.33

41 $1/t$, $1/3$

42 $x = 3\cos\theta - \cos 3\theta$, $y = 3\sin\theta - \sin 3\theta$

43 $(1, 0)$

Chapter 3

1 $v = \frac{1}{2}k\sqrt{t}$, $a = -\frac{1}{4}kt^{-3/2}$

2 12 m/s, -4 m/s^2

3 0, 6 m/s^2

4 $4\cos 2t - 12\sin 3t$, $-8\sin 2t - 36\cos 3t$

5 (a) $v = 6t$, (b) 6.9 m/s

6 (a) $v = 5 + 6t$, (b) 11.9 m/s

7 $\omega = 10 - 2t$, $\alpha = -2$

8 0.5 s

9 $\omega = \dfrac{r\omega_c\cos\theta}{L}$, $\alpha = -\dfrac{r\omega_c^2\sin\theta}{L}$

10 5.34 m/s at 68.2° to horizontal

11 14.1 m/s at 45° to horizontal

12 As given in the problem

13 As given in the problem

14 $-3m\sin t$

15 $i = -6\,e^{-3t}$ A

16 8 W

17 $16\cos 2t$ mA

18 $16\cos(4t + 45°)$ mA

19 $50\,e^{-100t}$ V

20 $(1 - 2t)e^{-2t}$ V

21 $v_L = Ve^{-Rt/L}$

22 $\dfrac{eAke^{-kx}}{(1 + e^{-kx})^2}$

23 (a) (0, 0), (b) $(-1/4, -1/4)$, (c) $(\pi/2, 1)$, $(3\pi/2, -1)$, etc., (d) $(\infty, 0)$

24 $y = u^2/2g$

25 (a) $(2, -8)$ minimum, $(-1, -19)$ maximum,
    (b) $(\pi/3, -0.68)$ minimum, $(5\pi/3, 6.97)$ maximum,
    (c) $(-3, 5)$ minimum, $(3, 1/5)$ maximum, (d) $(0, 1)$ minimum

26 $\frac{1}{2}(a - b)$

27 $x = L/2$

28 $R = r$, $V^2/4r$

29 20, 20

30 $h = 4$ m, $r = 2$ m

31 625 m^2

32 $6 \times 6 \times 3$ cm

33 707

34 47.7 V

35 $50\,\Omega,\ 50\,\Omega$

36 $(a+2b\theta)\delta\theta$

37 0.5%

38 6075 mm³

39 –1%

40 $v=-40\sin 2t$, $a=-80\cos 2t=-4x$

41 6 m/s, –4 m/s²

42 $\pm\sqrt{10}$ s

43 Proportional to cube of displacement

44 Proportional to fourth power of velocity

45 (a) $v=12-4t$, (b) 3 s

46 20 rad/s

47 $\alpha=-4t$

48 $v=-L\omega\sin\theta$, $a=-L\omega^2\cos\theta-L\alpha\sin\theta$

49 $-e^{-2t}$ A

50 (a) $i=Ca$, (b) $i=-CV\omega\sin\omega t$

51 $v=-30\ e^{-3t}$ mV

52 $v=8\ e^{-600t}-8\ e^{-200t}$ V

53 (a) (1/4, –1/8), (b) (0, 0), (c) (0, 2), (π/4, –2), etc., (d) (∞, 0)

54 (a) (0.2π, 5) maximum, (1.2π, –5) minimum,
   (b) (–2, –14) minimum, (1, 13) inflexion,
   (c) (–2, 8) minimum, (2, 8) minimum,
   (d) (–0.69, –0.61) maximum, (e) (0, 1) minimum

55 $\theta=a/2b$

56 6.25 m/s

57 As given in the problem

58 $\pm2\sqrt2$

59 (a) $v=50-10t$, (b) 125 m

60 20 mA/s, $1000t=0$, π, 2π, etc.

61 11.1 m, 9.0 m

62 $A/2$

63 As given in the problem

64 As given in the problem

65 447

66 $0.60L$

67 Cross river at an angle to reach bank 7.07 m down river

68 $2\pi\sqrt{\dfrac{T}{gp}}$

69 height = diameter

70 +0.5%

71 (a) ±3 mm^3, (b) ±2 mm^2

Chapter 4

1 $y = 0 + 2(x - 1) + (x - 1)^2$

2 $y = 3 + 3(x - 1) + (x - 1)^2 - 2(x - 1)^3$

3 3.6

4 10.75 m/s, 12.70 m/s, 14.70 m/s

5 4

6 1.75

7 1.0017, error −0.0017

8 0.883

9 2.0

10 −0.4162

11 9667 mm/s

12 0.5004, −0.2503

13 $y = 3x + 2x^2 - x^3$

14 $y = 2 + (x - 2) + 3(x - 2)^2 + 2(x - 2)^3$

15 0.333

16 5.440

17 (a) 1.1, (b) 1.0, (c) 1.0

18 (a) 0.9947, 1.0806, 1.0806, (b) 1.0788, 1.0806, 1.0806

19 −1.6815

20 14.60

21 As given in the problem

Chapter 5

1 (a) $x^3 + x^2 + C$, (b) $-\cos x + C$, (c) $\tan x + C$

2 $v = \int \dfrac{i}{C}\,dt$

3 $x = \int (u + at)\,dt$

4 +7.5 square units

5 0

6 4

7 (a) 4, (b) 3, (c) −3, (d) 5, (e) improper with no limit

8 (a) $\frac{x^4}{4}+C$, (b) $\frac{x^{-6}}{-6}+C$, (c) $\frac{x^{1.3}}{1.3}+C$, (d) $\frac{x^{4/3}}{4/3}+C$, (e) 1.5,

(f) 0.25, (g) improper with no limit

9 (a) $2x+C$, (b) $-\frac{1}{2}e^{-2x}+C$, (c) $-\frac{1}{3}\cos 3x+C$,

(d) $\frac{1}{4}\{\ln|\sin 4x|\}+C$, (e) $\ln x+C$, (f) $\frac{1}{2}\sinh 2x+C$,

(g) $\frac{1}{5}\{\ln|\operatorname{cosec} 5x-\cot 5x|\}+C$

10 (a) 10, (b) 1/2, (c) e^2-e^1, (d) $\ln 2-\ln 0$,

(e) improper with limit 1

11 $x=\frac{t^2}{2}+C$

12 $\int\frac{1}{N}\,dt=-kt$

13 $i=\int\frac{v}{L}\,dt$

14 $\sin x+C$

15 $x=\int(3+2t)\,dt$

16 8

17 0

18 (a) $\frac{x^{10}}{10}+C$, (b) $\frac{x^{-4}}{-4}+C$, (c) $\frac{x^{3/2}}{3/2}+C$, (d) $\frac{x^{1/2}}{1/2}+C$, (e) $7x+C$,

(f) $-\frac{1}{3}e^{-3x}+C$, (g) $2e^{x/2}+C$, (h) $\sin x+C$,

(i) $\frac{1}{4}\{\ln(\operatorname{cosec} 4x-\cot 4x)\}+C$, (j) $-\frac{1}{2}\cos 2x+C$,

(k) $\frac{1}{2}\sin 2x+C$, (l) $\frac{1}{2}\sinh 2x+C$

19 (a) 7/3, (b) −1, (c) 3/4, (d) −1/2, (e) improper with no limit

20 $\theta=\int cI\,dI$

21 $q=\int 3\sin 50t\,dt$

22 $x=\int(10t+3)\,dt$

Chapter 6

1 (a) $\frac{5}{2}x^2+C$, (b) $-\frac{2}{3}\cos 3x+C$, (c) $\frac{3}{4}e^{4x}+C$

2 $Q^2/2C$

3 (a) $\frac{1}{3}x^3+\frac{3}{2}x^2+x+C$, (b) $\frac{1}{2}x^2+\ln x+C$,

(c) $-\frac{1}{2}\cos 2x+\frac{1}{2}\sin 2x+C$, (d) e^x-e^x+C,

(e) $\frac{1}{3}x^3+2x^2+4x+C$, (f) $\frac{1}{12}x^3-\frac{1}{2}x+C$,

(g) $2x^{1/2}+\frac{4}{3}x^{3/2}+\frac{2}{5}x^{5/2}+C$

4 $x=2t+\frac{3}{2}t^2+C$

5 (a) $\frac{1}{2}\sin(2x+5)+C$, (b) $\frac{1}{16}(2x+4)^8+C$, (c) $\frac{1}{2}e^{2x-1}+C$,

   (d) $\frac{1}{24}(4x^2+1)^6+C$, (e) $\frac{1}{6}\ln(3x^2+2)+C$, (f) $-\frac{1}{12}(2-3x)^4+C$,

   (g) $\frac{2}{9}(x^3+2)^{3/2}+C$, (h) $\frac{1}{2}\ln(x^2+2x+2)+C$,

   (i) $2(x+2)^{1/2}\left[\frac{1}{5}(x+2)^2-\frac{4}{3}(x+2)+4\right]+C$, (j) $\ln(x^3+2)+C$,

   (k) $\ln(x+1)+C$, (l) $\ln(x^2+3x+2)+C$

6 $V=\dfrac{Q}{4\pi\varepsilon_r\varepsilon_0}\ln\left(\dfrac{D-a}{a}\right)$

7 $\ln(\theta-\theta_s)=t+C$, or $\theta=A\,e^{kt}+\theta_s$

8 (a) $\frac{1}{2}x+\frac{1}{4}\sin 2x+C$, (b) $-\frac{1}{6}\cos 3x-\frac{1}{2}\cos x+C$,

   (c) $\frac{3}{4}\tan 4x-\frac{3}{4}x+C$

9 (a) $\frac{1}{2}\sec^2 x+C$, (b) $\frac{1}{7}\cos^7 x-\frac{1}{5}\cos^5 x+C$,

   (c) $-\csc x-\sin x+C$, (d) $-\frac{1}{6}\cos^6 x+C$,

   (e) $\frac{1}{6}\sin^3 2x-\frac{1}{10}\sin^5 2x+C$, (f) $-\frac{1}{5}\cos^5 x+\frac{1}{7}\cos^7 x+C$,

   (g) $\frac{1}{16}x-\frac{1}{64}\sin 4x+\frac{1}{48}\sin^3 2x+C$

10 As given in the problem

11 $\pi/2$

12 (a) $\tan^{-1}x+C$, (b) $-\dfrac{\sqrt{9-x^2}}{x}-\sin^{-1}\frac{x}{3}+C$, (c) $\frac{1}{3}\tan^{-1}\frac{x}{3}+C$,

   (d) $\frac{9}{2}\sin^{-1}\frac{x}{3}+\frac{x}{2}\sqrt{9-x^2}+C$, (e) $\sin^{-1}\frac{x}{3}+C$, (f) $\tan^{-1}\frac{x}{2}+C$

13 (a) $\sqrt{1+x^2}+C$, (b) $2\cosh\frac{x}{3}+C$, (c) $\sinh\frac{x}{2}+\frac{x}{2}\sqrt{x^2+9}+C$,

   (d) $-\frac{1}{x}\sqrt{1+x^2}+C$, (e) $\frac{x}{2}\sqrt{x^2-9}-\frac{9}{2}\cosh^{-1}\frac{x}{3}+C$

14 (a) $\frac{1}{2}(x+2)\sqrt{x^2+4x-5}-\frac{9}{2}\ln\left(x+2+\sqrt{x^2+4x-5}\right)+C$,

   (b) $\ln\left(\sqrt{x^2-6x+10}+x-3\right)+C$

15 (a) $\ln\left(\dfrac{1+\tan\frac{x}{2}}{1-\tan\frac{x}{2}}\right)+C$, (b) $\dfrac{2}{\sqrt5}\tan^{-1}\left(\dfrac{\tan\frac{1}{2}x}{\sqrt5}\right)+C$

16 (a) $\pi/4$, (b) 1, (c) $\dfrac{\pi}{3}+\dfrac{\sqrt3}{2}$

17 (a) $\frac{1}{2}x^2\ln x-\frac{1}{4}x^2+C$, (b) $\frac{1}{2}xe^{2x}-\frac{1}{4}e^{2x}+C$,

   (c) $-\frac{1}{3}x^2e^{-3x}-\frac{2}{9}xe^{-3x}-\frac{2}{27}e^{-3x}+C$, (d) $-x\cos x+\sin x+C$,

   (e) $\frac{1}{2}\sec x\tan x+\frac{1}{2}\ln(\sec x+\tan x)+C$,

   (f) $x^3\sin x+3x^2\cos x-6x\sin x-6\cos x+C$,

   (g) $x\tan x+\ln(\cos x)+C$, (h) $-\frac{3}{2}x\cos 2x+\frac{3}{4}\sin 2x+C$,

   (i) $x\ln 4x-x+C$

18 (a) 0.718, (b) 8.715, (c) –0.25

19 (a) $\frac{1}{5}e^{2x}(\sin x + 2\cos x) + C$, (b) $\frac{1}{5}e^{x}(\sin 2x - 2\cos 2x) + C$

20 $\dfrac{2\omega\,e^{-a\pi/\omega}}{\omega^2 + a^2}$

21 (a) $\dfrac{5}{x-1} - \dfrac{4}{x-2}$, (b) $\dfrac{4}{x+1} - \dfrac{3}{x+2}$, (c) $\dfrac{3}{x+2} - \dfrac{2}{x-3}$,

(d) $-\dfrac{1}{x-1} + \dfrac{1}{x-2} + \dfrac{4}{(x-2)^2}$, (e) $\dfrac{2}{(x+1)^2} - \dfrac{1}{x+1} + \dfrac{1}{x-2}$,

(f) $\dfrac{2}{x-1} - \dfrac{2x+3}{x^2+x-1}$, (g) $1 - \dfrac{3}{x+4} + \dfrac{1}{x-2}$, (h) $3x - 7 + \dfrac{12}{x+2}$,

(i) $x + 6 - \dfrac{8}{x-2} + \dfrac{37}{x-4}$

22 (a) $13\ln(x-6) - 3\ln(x-2) + C$,

(b) $3\ln(x+1) - 2\ln(x-2) + C$, (c) $-\frac{3}{2}x + \frac{1}{4}\ln(3-2x) + C$,

(d) $\frac{1}{6}\ln(x-3) - \frac{1}{6}\ln(x+3) + C$,

(e) $\frac{1}{2}x^2 + 2x + \frac{2}{3}\ln(x-1) + \frac{5}{6}\ln(2x+1) + C$,

(f) $-\ln(x-1) + 3\ln(x+2) + \ln(2x+1) + C$,

(g) $\frac{2}{25}\ln(x+1) - \dfrac{1}{5(x+1)} - \frac{1}{25}\ln(x^2+4) - \frac{3}{25}\tan^{-1}\frac{1}{2}x + C$

(h) $-\frac{1}{2}\ln x + \ln(x-1) - \frac{1}{4}\ln(x^2+4) - \frac{1}{2}\tan^{-1}\frac{1}{2}x + C$

23 0.47

24 (a) $\frac{1}{3}x^6 + C$, (b) $-2\cos 2x + C$, (c) $5\ln x + C$,

(d) $\frac{1}{3}x^3 + 2x^2 - 2x + C$, (e) $2x - \frac{1}{5}e^{-5x} + C$,

(f) $-2\cos x - 4\sin x + C$, (g) $\tan x - x + C$,

(h) $-\frac{1}{7}\cos 7x + \frac{1}{3}\cos 3x + C$, (i) $-\frac{1}{16}\sin 8x + \frac{1}{4}\sin 2x + C$,

(j) $x - x^2 + \frac{1}{3}x^3 + C$, (k) $\frac{2}{3}(x-1)^{3/2} + C$, (l) $\frac{1}{2}e^{x^2} + C$,

(m) $\frac{5}{6}(1+x)^{1.2} + C$, (n) $-\dfrac{1}{1+x} + C$, (o) $\frac{2}{3}(3x-1)^5 + C$,

(p) $\frac{1}{2}(x^2+x)^2 + C$, (q) $2\sqrt{x} - 6\ln(3+\sqrt{x}) + C$,

(r) $-\frac{2}{3}(x+2)\sqrt{1-x} + C$, (s) $2\sin^{-1}\frac{1}{2}x + C$,

(t) $2\sin^{-1}\frac{1}{2}x + \frac{1}{2}x\sqrt{4-x^2} + C$, (u) $\dfrac{x}{2(1+x^2)} + \frac{1}{2}\tan^{-1}x + C$,

(v) $-\sqrt{4-x^2} + C$, (w) $\dfrac{2}{\sqrt{5}}\tan^{-1}\left(\dfrac{1}{\sqrt{5}}\tan\dfrac{x}{2}\right) + C$,

(x) $\dfrac{1}{\sqrt{3}}\ln\left(\dfrac{\tan\frac{1}{2}x + 2 - \sqrt{3}}{\tan\frac{1}{2}x + 3 + \sqrt{3}}\right) + C$,

(y) $\frac{1}{6}\tan^{-1}\left(\frac{13}{12}\tan\frac{1}{2}x + \frac{5}{12}\right) + C$,

(z) $\frac{1}{\sqrt{7}} \ln\left(\frac{4 - \sqrt{7} + 3\tan\frac{1}{2}x}{4 + \sqrt{7} + 3\tan\frac{1}{2}x}\right) + C$, (aa) $\frac{1}{3}x^3\ln x - \frac{1}{9}x^3 + C$,

(ab) $\frac{1}{2}x^2e^{2x} - \frac{1}{2}xe^{2x} + \frac{1}{4}e^{2x} + C$, (ac) $\frac{1}{9}e^{3x}(3x - 1) + C$,

(ad) $-x\cos 2x + \frac{1}{2}\sin 2x + C$, (ae) $-\frac{1}{x}\ln x - \frac{1}{x} + C$,

(af) $(1 - 2x^2)\cos 2x + 2x \sin 2x + C$, (ag) $\frac{1}{5}x^5\ln x - \frac{1}{25}x^5 + C$,

(ah) $\frac{2}{3}x^{3/2}\ln x - \frac{2}{3}x^{3/2} + C$, (ai) $\frac{1}{13}e^{3x}(2\sin 2x + 3\cos 2x) + C$,

(aj) $\frac{1}{29}e^{2x}(2\sin 5x - 5\cos 5x) + C$,

(ak) $\ln(x - 1) - \ln(x + 1) + C$, (al) $\frac{1}{2}\ln(x^2 - 1) + C$,

(am) $x + \ln(x - 2) - \ln(x + 2) + C$,

(an) $\frac{1}{2}x + \frac{1}{5}\ln(x - 1) - \frac{9}{20}\ln(2x + 3) + C$,

(ao) $-\frac{1}{x-2} + 4\ln(x + 2) + C$, (ap) $\ln(x + 3) - \frac{2}{x+3} + C$,

(aq) $\frac{1}{3}\ln(x + 2) + \frac{2}{3}\ln(x + 5) + C$,

(ar) $\frac{1}{3}\ln(x + 2) + \frac{2}{3}\ln(x + 1) + C$

25 $x = \frac{4}{3}t^3 + t^2 + C$

26 $y = \frac{1}{3}x^3 + x^2 + x + C$

27 As given in the problem

28 $4(1 + t)^{5/2} + C$

29 (a) $\frac{1}{12}$, (b) $\frac{1}{2}\pi$, (c) $\frac{1}{2}\ln 2 = 0.347$,

   (d) $\frac{1}{2}\left[\sqrt{2} + \ln(\sqrt{2} + 1)\right] = 1.148$, (e) $\frac{1}{2}(e^{-1} - e^{-4}) = 0.175$,

   (f) $\frac{9}{4}\pi$, (g) -0.5, (h) 4.575, (i) 0.2, (j) 0.311, (k) 0.0628,

   (l) 2.523

30 $\frac{2q}{r}\left(\frac{\pi}{2} - 1\right)$

31 1.296

32 $\frac{\omega}{a^2 + \omega^2}(e^{\pi a/\omega} + 1)$

33 $\frac{k}{a - b}[-\ln(a - x) + \ln(b - x)] + C$

Chapter 7

1 23/3 square units

2 (a) 5/6 square units, (b) –1/6 square units, (c) 2/3 square units

3 (a) 1 square unit, (b) –1 square unit, (c) 0

4 (a) 2/3 square units, (b) 18 square units, (c) 7/8 square units,

   (d) 2.296 square units

5 39 m

6 3 square units

7 1.83 square units

8 $ab\pi$

9 (a) 4/27 square units, (b) 1/12 square units, (c) 2 square units,
   (d) 9/2 square units

10 3.09 square units

11 (a) 2π cubic units, (b) 420.6π cubic units,
    (c) 318.4π cubic units, (d) $11\pi^2 + 24\pi$ cubic units

12 (a) 129.6π cubic units, (b) 40.5π cubic units

13 (a) 170.7π cubic units, (b) 9.6π cubic units,
    (c) 0.0571π cubic units

14 125.3π cm^3

15 $2\pi r$

16 (a) 12.295, (b) 6.042, (c) 0.881

17 $2a \sinh(d/a)$

18 (a) 4.59π square units, (b) 6.79π square units,
    (c) 83.46π square units

19 11.19π square units

20 $4\pi r^2$

21 609 square units

22 12 square units

23 0

24 4.05 square units

25 16.78 square units

26 12.3 m

27 (a) 4.5 square units, (b) 9 square units, (c) 4.5 square units,
    (d) 0.5 square units

28 3.62 square units

29 (a) 525π cubic units, (b) π cubic units, (c) $\pi/7$ cubic units,
    (d) 3.13 cubic units, (e) 10π cubic units

30 (a) 4π cubic units, (b) $\pi/5$ cubic units, (c) $2\pi/3$ cubic units

31 (a) 19.2π cubic units, (b) 85.3π cubic units

32 (a) 7.3 units, (b) 2.96 units, (c) 2.41 units, (d) 0.82 units

33 (a) 333.8π square units, (b) 27.16π square units,
    (c) 8.5π square units, (d) 82.39π square units

34 4.58π square units

35 $\pi r \sqrt{r^2 + h^2}$

Chapter 8

1 $4L/5$

2 2.57 m

3 8/3 along y-axis

4 126 mm above base of cylinder

5 185.7 mm above centre of base

6 (2, 4)

7 (2.31, 2.79)

8 ($\pi/2$, $\pi/8$)

9 (2/5, 4/7)

10 $\dfrac{4r\sqrt{2}}{3\pi}$ along centre of middle radius

11 (9/5, 9/5)

12 (1/2, 2/5)

13 (a) (20 mm, 20 mm), (b) (40 mm, 10 mm)

14 $\frac{1}{2}(R+r)\sqrt{h^2+(R-r)^2}$

15 $\sqrt{3}\,\pi L^2$, $\frac{1}{4}\pi L^3$

16 39.5 cubic units

17 74.7×10^3 mm^3

18 $\frac{1}{3}ML^2$, $\sqrt{\frac{1}{3}}L$

19 $\frac{4}{3}ML^2$, $\sqrt{\frac{4}{3}}L$

20 $\frac{3}{10}Mr^2$, $\sqrt{\frac{3}{10}}\,r$

21 $\frac{5}{4}Mr^2$

22 (a) $Mh^2/18$, (b) $Mh^2/6$

23 $2.28\pi\rho r^2$

24 0.0264 kg m^2

25 (a) $\dfrac{\pi}{4}r^4$, (b) $\dfrac{(9\pi^2-64)r^4}{72\pi}$, (c) $\dfrac{\pi(D^4-d^4)}{64}$

26 $\dfrac{BD^3-bd^3}{12}$

27 3.27×10^6 mm^4

28 60.3 mm

29 10.7×10^6 mm^4

30 $2L/3$

31 3.6 m

32 $3h/4$

33 34.4 mm above open end

34 (0, 1.6)

35 (9/8, 27/5)

36 (1.5, 3.6)

37 (3/2, 3/5)

38 (8/45, 2/3)

39 (2/5, 1/2)

40 (50 mm, 80 mm)

41 $4\sqrt{2}\,\pi L^2$, $\sqrt{2}\,\pi L^3$

42 $2\pi r L$, $\pi r^2 L$

43 7.24×10^6 mm^3

44 $\frac{1}{2}Mh^2$, $\sqrt{\frac{1}{2}}\,h$

45 $\frac{3}{10}Mr^2$, $\sqrt{\frac{3}{10}}\,r$

46 $\frac{5}{2}Mr^2$, $\sqrt{\frac{5}{2}}\,r$

47 $\frac{2}{3}ML^2$

48 $\frac{4}{3}ML^2$

49 $\frac{41}{38}Mr^2$

50 (a) $\frac{1}{36}bh^3$, (b) $\frac{1}{36}bh^3$, (c) $\frac{1}{36}hb^3$

51 40×10^6 mm^4

52 101×10^6 mm^4

53 5.03×10^6 mm^4

Chapter 9

1 (a) 7, (b) 17.3, (c) 8, (d) 2, (e) 1.56, (f) 3.91

2 $2A/\pi$

3 $20/\pi$

4 $\pi r/4$

5 $0.623N_0$

6 $\frac{1}{2}IV\cos\phi$

7 4.92

8 1.225

9 0.707

10 $V/2$

11 $I/\sqrt{3}$

12 (a) 24, (b) 5, (c) $10/\pi$, (d) 0.549, (e) $\frac{1}{\omega}(1-\cos\omega+\sin\omega)$, (f) 1

13 1.274

14 $\dfrac{2u^2}{\pi g}$

15 $20/\pi$

16 $8a/3$

17 $A/\sqrt{3}$

18 5.292

19 $\sqrt{\frac{1}{2}\left(I_1^2 + I_2^2\right)}$

20 $V/\sqrt{2}$

21 0.816 V

22 (a) 6.93, (b) 3.54, (c) 1.41, (d) 0.73, (e) 4.53

Chapter 10

1 (a) 2.255, (b) 2.326

2 (a) 0.791, (b) 0.787

3 0.7471

4 (a) 0.6970, (b) 0.5123, (c) 1.006, (d) 4.063, (e) 33.328

5 0.935 J

6 0.55 C

7 (a) 0.511, (b) 3.196, (c) 1.568, (d) 0.905, (e) 0.785, (f) 1.187,
(g) 10.070, (h) 0.648

8 6.391 J

9 9.27 m

10 25.73 m^2

11 Analytical 0.7854, (a) 0.7828, error 0.0028,
(b) 0.7854, error 0

12 Analytical 2.5452, (a) 2.5297, error 0.0155,
(b) 2.5447, error 0.0005

13 2.000

14 0.746 824

15 (a) 1.108, (b) 25.25, (c) 0.799

16 (a) 2.750, (b) 0.509, (c) 1.502, (d) 0.681, (e) 0.647

17 7.78 mC

18 (a) 6.391, (b) 0.372, (c) 1.571, (d) 0.186, (e) 0.479, (f) 0.246,
(g) 2.927

19 4.67

20 Analytical 4.0000, (a) 4.2500, error −0.2500,
(b) 4.0000, error 0

21 Analytical 6.0000, (a) 6.01313, error −0.01313,
(b) 6.0000, error 0

22 2.545

23 0.2929

24 0.7500

25 0.5108

Index